Waiting for the Big One

Charlotte Mazel-Cabasse

Waiting for the Big One

Risk, Science, Experience, and Culture in Disaster Preparedness

palgrave
macmillan

Charlotte Mazel-Cabasse
Data Science
Berkeley Institute for Data Science
Berkeley, CA, USA

ISBN 978-3-030-15288-8 ISBN 978-3-030-15289-5 (eBook)
https://doi.org/10.1007/978-3-030-15289-5

Cover images: koksikoks and Somsak Bumroongwong/EyeEm
Cover design by Tjaša Krivec

This Palgrave Macmillan imprint is published by the registered company Springer Nature Switzerland AG
The registered company address is: Gewerbestrasse 11, 6330 Cham, Switzerland

For Etienne, Simon, Antonin,
and for my parents.

Preface

Disasters impact millions of people every year. Some of these catastrophes gain high visibility in international media and some do not. When they do, reporting tends to focus on the sequence of events that "make" the disaster: perimeter, casualties, and cost. Sometimes, media become interested in what happened before (often in a search for responsibilities) and after. But in the life of the people affected by a disaster, these two phases—preparedness and mitigation—are crucial and can transform the entire experience of going through catastrophic events.

In risk-prone areas, experts and concerned residents take the burden of considering how to best prepare for disasters. In the San Francisco Bay Area, they have called themselves the Earthquake Junkies, playing with the idea that preventing risk is like an addiction, something from which they cannot desist, despite knowing that it may be detrimental to their own health and sanity. As I learned while undertaking this research, being an Earthquake Junkie is hard: it requires understanding the mechanics of the Earth's crust and the dynamics of social environments. It requires the building of bridges between the hard and the soft (Serres, 2006), scientific rigor and humility, and being able to reconcile the many ways in which we—collectively—think about a disaster and can help mitigate it. It also requires what French philosopher Emilie Hache has described as a too-often disregarded competence of the Moderns—that is, the capacity to use our emotional response to disaster as the trigger for constructive action to "collectively put words on a collective fear and draw energy to act" (quoted in Vincent, 2017).

In this book, I have tried to show that disaster (such as earthquake, but also fire) has several existences, they can—at the same time—be a memory, a projection, and a probability. They are the sum of existing knowledge of their mechanism at a moment T, but they can also be a trauma, a source of jokes, an existence you want to forget. When I began this field research, the prevailing concept within the community of scientists and experts was that expert and lay knowledge were of different natures and could not be compared, one being so evidently superior to the other. I faced brutal pushback when I started talking about the emotions and life experience of experts. Experts and scientists were not supposed to have emotions; science was factual and rational, not emotional.

During the many years that I have spent in California, as a resident, a student and, later, as a researcher, this conception of the knowledge has been challenged time and time again. Not only did the experts that I met during this research have an appreciation that was richer and more complex than I had ever expected, but they also paid attention to the interwoven and hybrid nature of the knowledge that they were producing. Experiencing the risk every day and working against it gave them a nuanced—although not naïve—understanding of what it is to cherish a place, its people, its culture and lifestyle and do everything to make sure that a disaster will not reduce it to ashes.

The experts that I met might have been special in some ways: the Bay Area metropolitan area was definitely something that residents, new and old, earthquake experts or not, cared deeply about; the experience of the Oakland Fire—forgotten by newcomer but still vivid for those who had had to live through it—kept people on the alert. But I found that experts were driven by the cheerful understanding that all of us, as a community, will live better lives if some take on the burden of doing the risk preparedness work, knowing that others were taking on the nonetheless important aspects of collective life.

For a long time, I thought that Earthquake Junkies would be very similar to climate scientists. They both developed incredible knowledge about a system that could go wrong; they take it upon themselves to educate the rest of us and ensure that the disaster that was certain to happen could be mitigated in some way. They were on a mission, worked long hours to improve the science, and tried to make decisions with limited resources. As the COP24 (December 3–14, 2018, in Katowice, Poland) shows, I realize this was perhaps an optimistic comparison. While, in the Bay Area, earthquake experts had been listened to and their advice and recommendations—although not followed to the letter—have found their way to policy-makers and residents, matters seem to be taking a very different trajectory when it comes to climate change.

To be useful, Earthquake Junkies need to be heard and their contribution recognized. As Isabelle Stengers urges us, protecting the common good is a collective action and we need to become better at the "art of paying attention". As she pointed out: "If there is an art and not just a capacity, this is because it is a matter of learning and cultivating, that is to say, making ourselves pay attention. Making in the sense that attention here is not related to that which is defined as a priori worthy of attention, but as something that creates an obligation to imagine, to check, to envisage, consequences that bring into play connections between what we are in the habit of keeping separate. In short, making ourselves pay attention in the sense that attention requires knowing how to resist the temptation to separate what must be taken into account and what may be neglected" (Stengers, 2015: 62).

Berkeley, USA — Charlotte Mazel-Cabasse

References

Serres, M. (2006). *Homo pontifex. L'art des ponts*. Paris: Le Pommier.

Stengers, I. (2015). *In Catastrophic times: Resisting the coming barbarism* (Critical C). London, UK: Open Humanities Press and meson press. Retrieved from http://openhumanitiespress.org/books/download/Stengers_2015_In-Catastrophic-Times.pdf.

Vincent, C. (2017, December 29). En 2018, vous ne verrez plus le réchauffement climatique comme une fatalité grâce à la philosophe Emilie Hache. *Le Monde*. Retrieved from http://www.lemonde.fr/idees/article/2017/12/29/en-2018-vous-ne-verrez-plus-le-rechauffement-climatique-comme-une-fatalite-grace-a-la-philosophe-emilie-hache_5235552_3232.html.

Acknowledgements

I had the privilege to nurture this project with a fantastic editorial team at Palgrave Macmillan and Springer Nature: I thank my editors Rachel Daniel, Kyra Saniewski, and Madison Allums, as well as the fantastic production team: Azarudeen Ahamed Sheriff, Prathipa Raju and Brian Halm for their dedication and patient work throughout the different stages of publication.

This book owes much to a group of dedicated scholars that organized workshops and collaborations across the globe: Professor Joonhong Ahn, who unfortunately passed away in June 2016, and whose contribution to the field has been of incredible importance; Cathryn Carson, who helped me bridge different interests and fields of research in a way for I shall always be grateful; Kim Fortun, Franck Guarnieri and Sulfikar Amir.

Patiently following my intellectual travels and geographical wanderings (or was it the other way around?), Valérie November helped me not to deviate from the course of what graduate school is supposed to be. I would also like to thank Professors Saul Perlmutter, David Culler, and Henry Brady, as well as the Berkeley Institute for Data Science's outstanding team of fellows and co-PIs and the program directors of The Gordon and Betty Moore Foundation and The Alfred P. Sloan Foundation for helping me embark on other fascinating—and risky—projects.

It has been a great pleasure to conduct interviews with the scientists and experts regarding earthquake risk in the Bay Area. Here, I specifically want to express my gratitude to Marjorie Green, Special Projects

Manager at the Earthquake Engineering Research Institute (EERI), for connecting me with her network of colleagues and friends: the Earthquake Junkies. Each interview has opened a window on the many ways in which people have engaged with the earthquake in the Bay Area. I also want to thank Peggy Hellweg, from the Berkeley Seismological Laboratory, for helping me walk through the complexity of this new field of research.

Throughout the years of my study, the friendship of fellow researchers Hélène Mialet, Kyoko Sato, Reiko Hasegawa, Laura Noren, Stuart Geiger, Nelle Varoquaux and Christine Fassert, and Nicolas Benvegnu (to name a fews) has been a precious gift. I also want to thank my grandmother, my parents, my always inspirational siblings and cousins, and their growing families. Thank you also to my extended family for being a constant support in our lives. My dear friends have been always patient and encouraging: thank you for bearing with me, with us, all these years. This book could have not been written without the dedication of a team of babysitters and nannies who extended my working days and have been infinitely patient (or so it seems) with our messy organization. It has also beneficated of the careful editing of Donald L. Woodward, Kristine Jones and Karen Kelsky (@The Professor Is In). As for my own family: I owe you more than I can say, and I am looking forward to starting a new chapter of our life together.

This work has beneficiated from the support of the Swiss National Science Foundation.

Contents

List of Figures

CHAPTER 1

Introduction

On March 11, 2011, the Tōhoku earthquake and tsunami, also referred to in Japan as the Great East Japan Earthquake, partially destroyed the Fukushima Daiichi Nuclear Power Plant. These events caused the deaths of 15,884 people, led to the evacuation of 300,000 others, and triggered a nuclear accident the causes and long-term consequences of which are still under investigation (Ahn et al., 2015; Guarnieri & Travadel, 2018; Hasegawa, 2013; Kalmbach, 2015). The series of events constituting the earthquake and tsunami, and the scale and amplitude of the ensuing nuclear disaster, were breathtaking and took the international community by surprise.

In Berkeley, California, the emotions aroused by the catastrophe and the threat of a nuclear disaster put residents on the alert. Like millions of others, I was glued to my computer—watching CNN live, trying to make sense of the information I had, and speculating on what was not yet known. Everywhere—in supermarkets, in playgrounds, at work—discussions revolved around the disaster, the sorrow, the pain, and the risk. On April 20, 2011, an interdisciplinary group of faculty members from the University of California, Berkeley (UC Berkeley), gathered for an open discussion entitled "Coping with the Crisis: Implications for Japan's Future."[i] That evening, the panelists openly and genuinely shared their thoughts and their knowledge of what had happened—what it meant not only for the affected communities, but also for rest of us. The room was full, faces were grave; the discussion continued for many hours as people tried to sort out the information coming from diverse

C. Mazel-Cabasse, *Waiting for the Big One*,
https://doi.org/10.1007/978-3-030-15289-5_1

sources: Japanese government, news agencies, citizen science network, scientists (Shineha & Tanaka, 2017).

Residents, faculty members, and concerned citizens were wondering whom to trust and how to move forward from this point.[ii] This discussion was the first of many organized at UC Berkeley, generated by a group of dedicated and concerned scholars[iii] willing to use their knowledge and energy to limit the spectrum of the catastrophe unfolding before our eyes, and to inform the public and the policy-makers. Relying on an international network of experts, citizen scientists, academics, friends, and family members, they translated and discussed information, weighing contributions from those in a position to take part in this large enterprise of interdisciplinary sense-making. As a graduate student working on the earthquake risk in the San Francisco Bay Area, I was invited to participate in three of these workshops (Ahn, Guarnieri, & Furuta, 2017; Akera, 2007; Amir, 2018), where scholars tried to find a common language to describe the complexity of the disaster that had been such a deep emotional and intellectual shock. Building on what French philosopher Emilie Hache has described as a disregarded competence of the Moderns: the capacity to use our emotional response as the trigger for action to "collectively put words on a collective fear and draw energy to act" (quoted in Vincent, 2017) people in the room were joining forces to think through the multiple, and often contradictory dimensions of disasters on the scale of the Fukushima Daiichi event.

1.1 A Questioning Situation

Studying an event of the scope of the Tōhoku earthquake, tsunami, and ensuing nuclear disaster with the tools of academic knowledge is a humbling experience: no single discipline, method, approach, or scholar can either describe or explain the chain of reactions leading to a disaster of that scale.

As the discussions in the aftermath of the incident made explicit, the risk of a catastrophic event is thought of as an association of distinctive modalities, or modes of knowing, that crosses the traditional divisions of academic disciplines and methods: "[D]isasters come into existence in both the material and the social world and perhaps, in some hybrid space between them" (Oliver-Smith & Hoffman, 1999: 24). What seems coherent and valid from the perspective of the event is sometimes hard to articulate and prone to debate from the perspective of academic disciplines. To account for this complexity, anthropologists have argued that disasters "are both socially constructed and experienced differently by different groups and individuals, generating multiple interpretations of an event process. A single disaster can fragment into different and conflicting sets of circumstances and interpretations according to the experience and identity of those affected" (Oliver-Smith & Hoffman, 1999: 26).

What resources do we have to approach such a catastrophe? How can we deal with the different perspectives, meanings, and consequences of what seems at first a single event but whose ramifications and multiple specificities can easily overcome our capacity to think? How useful is the knowledge of past disasters in addressing new ones? Like all other disasters, the Fukushima disaster had multiple *existences*, multiple ways of "being in the world," that were hard to reconcile. This apparent incompatibility seemed reinforced by the many ways in which the narration, the stories, and the analysis of the event were performed, across disciplinary fields and epistemologies.

1.2 The Uncertain Space of Risk

My field research—localized in the Bay Area of San Francisco, California, on the West Coast of the United States—seeks to explain how the possible displacement of two massive tectonic plates along the Pacific Rim could become "the Big One," the next major earthquake—an event both feared and awaited, and yet invisible. The West Coast of the United Sates is particularly concerned by the possibility of large-scale earthquakes. In and around the Pacific Ocean, the junction of the Pacific plate with other plates (the North American plate, Cocos plate, Juan de Fuca plate, Nazca plate, Antarctic plate, Australian plate, and Filipino plate) creates a very active seismic zone, commonly called the "Ring of Fire." In the past, other earthquakes have also been called a Big One: the M6.8 1868 Hayward earthquake was, for instance, a Big One before being dethroned by the M7.8 1906 San Francisco earthquake. The most recent earthquakes, however—the Loma Prieta earthquake in 1989 and the M6.0 2014 South Napa Earthquake that hit the city of American Canyon to the north of the San Francisco Bay Area—were never considered sufficiently strong to qualify for the name.

1.2.1 *The Multiple Existences of the Earthquake*

Risk and disaster are hard to characterize fully, and they resist the observer's temptation to determine a single definition or a simple explanation. However, the manner in which experts, scientists, residents, and the public comprehend them has a considerable impact on the politics of mitigation and reconstruction. Since the end of the Second World War, a long tradition of research has explored and critically analyzed the intertwined relationships between micro- and macro-level events, and the multiple stakeholders involved in the unfolding of disasters (Boudia & Jas, 2007; Frickel & Bess, 2007; Hoffman & Oliver-Smith, 2002; Knowles, 2011; Lane et al., 2010; Lash, Szerszynski, & Wynne, 1996; Quarantelli, 1998; Solnit, 2009; Tierney, 2001; White, 1945; Wisner, Blaikie, Cannon, & Davis, 2006). Over the past few decades, qualitative and quantitative approaches have continued to coexist, creating different forms of sociotechnical assemblages pursuing the same objective: obtaining a more precise representation of earthquake signals and generating a clearer interpretation of the mechanism that triggers the movement of tectonic plates.

In seismology and earthquake studies, long before the era of the constant monitoring of activity (Bossu et al., 2011; Taira, Silver, Niu, & Nadeau, 2009), knowledge of earthquake risk came from first-hand experience and the evaluation of damage in the immediate aftermath of a catastrophe. Seismologists were therefore required to show some awareness of the sociotechnical context in which the disaster happened blurring the categories of expertise and lay knowledge, civic and scientific engagement. Building on years of field work and research, experts scientists used their intimate knowledge of the earthquake—how it felt, how it displaced things, how it changed the landscape—as a topic for scientific inquiry. This multi-dimensional knowledge of the earthquake has prompted scientists to consider the experience of living through earthquakes as a valid subject for scientific study, while motivating residents to better educate themselves.

An important part of the historical developments of seismology took place in California and more specifically in the Bay Area in the early twentieth century. The field went through an transformation following the 1906 San Francisco earthquake and fire when data collection became more systematic and organized. In the aftermath of the event, an extraordinary expedition was set up to document the consequences of the earthquake through a State Earthquake Investigation Commission order by

Governor George C. Pardee and lead by geologist Andrew Lawson. The Lawson Commission's report that resulted was the first full-scale attempt at comprehensively documenting an earthquake. Rupturing 296 miles of the San Andreas Fault, the magnitude 7.9 earthquake indeed "afforded an exceptional opportunity for adding to our knowledge of earthquakes," noted Lawson, who was also the chair of the Department of Geology at UC Berkeley (in Lewis, 2008). To accomplish this prodigious task, he had to dispatch teams of observers on foot and horseback to explore the Fault from Humboldt County in northern California to the Coachella Valley, south of Los Angeles and, by 1908, they had mapped almost the entire San Andreas Fault. The ensuing report included the elastic-rebound theory, a important step in explaining earthquake mechanism.[1]

From that moment, interest in earthquakes as phenomena took off in California. During the early part of the twentieth century, Harry O. Wood, Franck Neumann, and Charles Richter defined scales of intensity and magnitude. These were conceived as interpretive frameworks: tools aimed at describing earthquakes in words and situating them on a scale. In 1931, Harry O. Wood—who had been working for decades with eyewitness earthquake observation reports (also known as "felt reports")—published the Modified Mercalli intensity scale with Franck Neumann. This new scale was designed to make reporting easier by defining an earthquake as having degrees and thresholds, thus eliminating ambiguities, but also to "insert explicit statement[s] about the mental states conducive to certain reported effects" (Coen, 2013: 258).[2] The Richter magnitude scale, which measures the strength of an earthquake, was then introduced in 1935. By the 1960s, only after tectonic plate theory[3] had largely been accepted by the scientific community seismologists were

[1]According to these theories, earthquakes were due to the sudden release of strain that had been gradually built up by the constant creeping of the earth's surface near a fault. In his contribution to the commission's final report, Harry Fielding Reid had argued that there has indeed been a gradual distortion of the earth's surface near the San Andreas Fault during the late nineteen century, just as the elastic rebound theory called for (Geschwind, 2001: 60–61).

[2]The Modified Mercalli Intensity Scale is still used today in the Shake Map, also known as the "Did You Feel It?" map.

[3]The theory of tectonic plates, formulated between 1956 and 1967 (Frankel, 2012), validated that earthquakes result from movements created by the accumulation of energy in the always-moving Earth's crust.

able to describe, and sometime explain, the mechanisms that trigger an earthquake.

Despite years of research, earthquakes still surprise (Lussault, 2007), sometime even decades after their occurrence. For instance, the scale of the great 1906 earthquake and fire has been difficult to appreciate. A century after the fact, reopening a seemingly cold case, Stephen Tobriner, Professor Emeritus of Architectural History in the Architecture Department at the University of California, Berkeley, showed that—contrary to the commonly accepted narrative—the 1906 earthquake did little damage when compared to the fire that followed[4] (Tobriner, 2006). The discussion about the nature of the disaster—whether it had been caused by the earthquake or the fire—and its consequences has been of prime importance because of its political and economic implications (Brechin, 2006; Walker, 1990). But it also matters for today's science as observations of past catastrophes have been (Coen, 2013), and still are (Juraku, 2015; November & Leanza, 2015), a major source of information for scientists and experts engaged in disaster preparedness.[5].

1.2.2 A Word About the Position of the Researcher

A Bay Area resident between 2008 and 2018, I have been embracing my field research fully, so much so that I have called this region "home." Living in the East Bay, playing a part in its urban environment, I have known that my family and I have never been further than 500 meters from the Hayward Fault. Like all other residents, I have had to make decisions and evaluate earthquake risks for my family and for myself.

[4]Looking at pictures taken directly after the earthquake but before the fire, Tobriner established a new estimation of the 1906 destruction. In doing so, he also contested what "many scholars and popular historians have accepted and repeated, [namely] the idea that San Francisco of the 1860s denied the existence of seismic danger [...]. However, historical records show that architects, engineers, and even everyday citizens understood the consequences of the earthquake of the 1860s and tried to inventory the damages, to understand what had happened, retrofit buildings to resist future earthquakes and to build earthquake resistant structures" (Tobriner, 2006: 35).

[5]In 2006, scientists and experts modeled and analyzed the possible impacts of an earthquake similar to the 1906 Big One in today's San Francisco (Perkins et al., 2006). The multifaceted dimensions of the earthquake as a phenomenon emerging progressively as a measurable object are traceable until today. If the earthquake contours have remained—and still are—difficult to grasp fully, this (partially) undefined state has allowed the multiple and concomitant forms of understanding of the phenomenon co-exists.

Our first rented apartment was located in the Berkeley Hills: a large un-retrofitted three-story building planted on the hillside, surrounded by large trees, with only one winding road for access and a stunning view of the Golden Gate Bridge. I experienced my first earthquake in this house; this is also where my son was born in 2010. While doing interviews for this research, I learned that: (1) hillside houses had been responsible for most of the casualties in Los Angeles during the 1994 Northridge earthquake (Von Winterfieldt, Roselund, & Kisuse, 2000), and (2) most of the casualties during the 1991 fire occurred because of traffic congestion in the very narrow Oakland Hills roads; people had been trapped in their cars when the fire reached them. In addition to living in an unsafe house in a fire hazard area, it so happened that my son's first preschool classroom was in the basement of a 1950s building with concrete walls 20 meters high. From my window, I could see the Oakland Hills properties that had since been rebuilt. I heard the stories of those who had been through the disaster—those who died, those who escaped, those who decided to leave, and those who rebuilt.

As local residents, we had to adapt to the extreme tightening of the rental market. As parents, we also had to think about the building in which our children were spending their daytime. Was it safe, has it been properly retrofitted? Were the teachers and school staff trained? Public schools have since developed mandatory procedures to deal with an emergency, but this was not the case at the preschool that my son attended for a while. Over time, and after some very dark thoughts, nightmares, and cold sweats when I dropped him off, I finally found the courage to talk to the teachers, alerting the school to the dangers of an un-retrofitted building. The teachers themselves felt unsafe in the building and the school board took the matter seriously enough to decide to carry out a certain amount of retrofitting work. We eventually moved to a less hilly place to live in a retrofitted house.

Between 2009 and 2013, I interviewed earthquake experts (seismologists, building inspectors, urban planners, structural engineers, and others), focusing on the kind of people Tobriner referred to when paying them a tribute[6]: "They are us: the homeowner making his home safe for his family, the investor building apartments that are safe for his family, the businessman who pays extra for earthquake safety, the inventor who

[6]Excluded from the interview process was the insurance and actuarial sector, which is covered by other ongoing research (Johnson, 2011).

deals with innumerable vexing details, the research scientist trying to understand earthquakes, the government official who implements new earthquake policies, the engineer concerned about earthquake safety who does his best to ensure it; the architect who realizes that safety is as important as architectural expression; the contractors and subcontractors who understand earthquake safety and build for it" (Tobriner, 2006: 280). Many of my interviewees did, in fact, overlap in more than one category. During these interviews, it became evident that these individuals form an epistemic community (Knorr-Cetina, 1999, Mahony & Hulme, 2016) enmeshed with the history of the development of resilient infrastructures. Their personal histories, professional trajectories, and epistemic networks have defined the contours of this sociotechnical assemblage, this network of awareness to risk—from which emerged the scientific definition of earthquake risk and the possibility of resilience.

While listening to my respondents, it occurred to me that mundane decisions (where to live, which school to choose, how to travel) were never studied and considered alongside more academic approaches. Scientists and experts would never be residents, parents, and commuters and would not be grounded in any kind of everyday, mundane life. But, very rapidly, the first interviews that I conducted revealed more than the ungrounded, know-it-all attitude associated with expertise. My respondents were questioning their knowledge and their policy; they listened, they appreciated questions and welcomed feed-back. As I progressed with my data collection, I found myself dealing with material that had no evident "place" within the corpus of social science research in disaster studies. For example, I was left wondering how best to incorporate the emotions of a building inspector as she revealed her post-traumatic stress disorder (PTSD) after being commissioned to evaluate the damage of houses destroyed during the Northridge earthquake in which ten residents had perished. And how much should I include of an other UC Berkeley architecture professor crying at the memory of her lost home, destroyed in the 1991 Oakland Fire?

As I pursued my field research, it became clearer that the subjective dimensions of risk should be not be removed from my work, but, rather, that they should be taken seriously: they were not "irrational" perceptions or emotions but, rather, actants, in the Actor-Network Theory (ANT) sense. For these experts and residents a disaster is not something that does only happen to others: the same fears and preparatory measures occur where they live and work, as well.

1.2.3 *Actants, Agency, and Assemblages*

Looking for the footprints of risk, I employ the dynamic understanding of the spatiality of risk, seeking not only stories that both recall the event and redefine the place (Lamy, 2007), emotions (Bondi, 2005; Davidson, Bondi, & Smith, 2005; Thien, 2005) and that shape the everyday practices of space (Anderson & Harrison, 2010; de Certeau, 1988), but also the scientific and urban projects that define the specificity of the Bay Area (Lévy & Lussault, 2000). A non-spatialized risk is an abstract threat, as much as space without any risks—which would be like a base map—remains an object of fiction, an imagined territory that bears few similarities with the materiality of what makes space (November, Camacho-Hübner, & Latour, 2010). On the other hand, anticipating a disaster is a complex operation which includes a need for the capacity to organize various forms of knowledge, temporalities, and spatial dimensions. Being a resident of the Bay Area implies being an observer of your own environment as well as facing the daily risk of eruption, requiring us to question how to "deal with" an earthquake and to "dwell" in an unstable environment (Breviglieri, 2006; Lazzarotti, 2006; Paquot, Lussalut, & Younes, 2007; Stock, 2004).

As an extension of this scientific tradition of co-construction, Tobriner had identified what I would argue to be a key element of this socio-technical infrastructure: the collective network of individuals, citizen observers, and citizen scientists who pay attention to the risk; who look for it and prepare for it. Tobriner exposed the social network of intertwined emotional and personal connections between people, artefacts, and places. Described as a specific form of "attachment" by (Gomart & Hennion, 1999), this network explains why residents dedicate lifelong careers to the mitigation of earthquake risk while simultaneously participating in the development of a thriving environment in the region to which they are so committed. Looking at the everyday existences of the earthquake, it describes how a group of concerned individuals have managed to navigate the different existences of earthquake risk, combining the result of extensive field research among earthquake experts, scientists, and concerned residents of the Bay Area.

This work brings light to a network of risk awareness, while *waiting for the Big One*. It does not intend to measure, or describe, the public's recognition of earthquake risk. Rather, I argue that *waiting for the Big One* is a collective experience that brings together both experts and non-experts in collaboration to mobilize residents' attention. By documenting the experience of a group of experts and scientists who call themselves "the Earthquake Junkies," it provides a perspective on a set of operations which, from residential choices to seismic-mapping activities, from memorialization to projection, define and help make the potential consequences of an earthquake visible.

In order to collect material for this research, I contacted 65 people and conducted 34 extensive interviews—most of them lasting for several hours. Having people switch between their professional and their personal experiences was not always easy: 3 interviewees clearly expressed their disapproval of being asked personal questions. However, most of the interviewees had little difficulty slipping between personal and professional accounts, acknowledging—often, even before I broached the subject—the open secret of the permeable frontier between hard science and personal experience.

I have chosen to focus on people locally engaged in risk preparedness who are also residents of the Bay Area.[7] The insights of these rather understudied categories of actors are unique in the field of risk and disaster studies. In building this sample, I have also tried to interview individuals whose professional activities have situated them at different temporal stages of an unfolding earthquake event. From the geological and mechanical dimensions of the phenomenon, to the questions it poses to structural engineers monitoring buildings and policy analysts crafting regulations, to first responders dealing with a catastrophe.

I have also focused on representatives of the Bay Area Earthquake Alliance, an organization that brings together the major actors of earthquake risk prevention: the California Emergency Management Agency (Calema), Association of Bay Area Government, California Geological Survey, California Seismic Commission, Berkeley Seismological Lab, Earthquake Engineering Research Center, and the American Red Cross. Along with these major actors, I interviewed city officials from the cities of Oakland, Berkeley, and San Francisco; urban planners and architects in charge of the reconstruction of Oakland after the Loma Prieta Earthquake; and independent researchers and lobbyists dedicated to the cause of earthquake safety in local, state, and national arenas. Finally, I have interviewed grassroots associations, which often had a different

[7]To the exclusion of the insurance and actuarial sector, which is covered by other ongoing researches (Johnson, 2011).

perspective on ideas concerning risk and disaster community management. They spoke for those who are not heard in this research: the most fragile and vulnerable populations toward whom these organizations are geared.

To guarantee diversity when it came to the varied experiences of the individuals discussing earthquakes, I identified categories that reflect the dominant type of relationship that people have established with the question of earthquake risk. After some interviews with residents with no particular interest in, or experience of earthquake risks,[8] I chose to reframe the focus on people who have, in one way or another, engaged with the risk of earthquake. In order to maintain confidentiality, the interviewees' full names are not disclosed in this work. This field research was also completed by reading the grey literature on the subject, thanks to the help of the librarians at the Pacific Earthquake and Engineering Resource Center Library on the UC Berkeley campus.

The first category comprises the scientists and researchers who have contributed to the creation of knowledge and the production of data. Here, again, they come from different backgrounds. I attempted to cover some of the connections regarding the phenomena of earthquakes: geophysicists, seismologists, computational engineers, architects, public scientists, geographers, structural engineers, and urban planners. They all work to produce data and, over the last few decades, some of them have taken their roles as spokespeople regarding earthquake risk very seriously, advocating for the improvement of infrastructure for better safety, as well as promoting more active engagement from the general public in risk preparedness.

People who had personal experiences of disasters comprise the second group of interviewees. These are the people who care and are concerned about earthquakes but have no professional training in this area. In this category, I interviewed people who had lost their homes during the 1991 Oakland Fire. The question of the integration of their testimonies, which follow the questionnaire, was important in the spectrum of voices heard for this research. The recollection of this massive, yet geographically limited disaster provides an important perspective regarding the questions of loss, resilience, and (re-)construction of a space of risk and disaster that has faded over time.

[8]This last category is represented to a lesser extent than the others, since it seems that the lack of interest in the question produced fewer interesting results for our specific research.

To this typology, we could add a third category: the media. This group is composed of writers, journalists, and essayists—some of them quite polemical—who have taken on the issue of earthquakes, or natural disasters more generally, in various publications and broadcasts. They form a very disparate aggregate—people with different backgrounds and interests, and very different perspectives on the question of California's natural disasters. Their contributions to the debate are important, as they can solidify certain common views about the questions I address here.

1.3 Paying Attention to the Process of Instauration

Looking at the study of sciences, historians noted that nineteenth-century scientists had "take[n] out the human element from the research, to make the research processes and products objective" (Strong, 2008). In doing this, they made invisible the multiple agency, the actants, the mode of existence—which "interfere" with the scientific process—and, in that same movement, took away the complexity of the subjectivity of the scientist as a knowing subject (Houdart, 2008; Mialet, 2012). Following the steps of Alfred North Whitehead, who contested what he called this "bifurcation of nature" (Latour, 1991, 2011),[9] contemporary researchers have looked at the hybrid "assemblages" of primary and secondary qualities, fact and rationality or irrationality, human and non-human (Datson & Galison, 2010; Fressoz, 2008; Houdart & Thiery, 2011; Latour, 1991).

In the second half of the twentieth century, the French philosopher Etienne Souriau (1892–1979) explored the possibility of *existential pluralism* and proposed that existences be perceived as multiple and coexisting modes, allowing us to describe associations or phenomena that are situated without being ethno- or geo-centered. Before him, and focusing

[9] This bifurcation begins somewhere between Galileo and Locke and comes to an end, in Whitehead's opinion, with William James. This brief period, which I call 'the modernist parenthesis'—during which we thought we were modern—has three main characteristics: the conviction that the world can be divided into primary and secondary qualities (which can be called 'naturalism'); the ever increased intermingling, in ever larger assemblages, of these same primary and secondary qualities (which can be called 'hybrids'); and lastly, a watertight division between the constantly repeated assertion that the division between primary and secondary qualities must be maintained, and the practical reality which is in fact the exact opposite of this theory (which one could call the 'obscurantism of the Enlightenment') (Latour, 2011: 1–2).

especially on earthquakes, the American philosopher William James, writing about his experience of the 1906 earthquake and fire that partially destroyed San Francisco, spoke of the articulations that define—for scientists, residents, and himself—the multiple but simultaneous existences of the earthquake. In a piece published under the title "On some mental effects of the earthquake," he reflects on the definition of the earthquake as an event that "wanted simply to manifest the full meaning of its *name*." He wrote, "For '*science*,' when the tensions in the earth's crust reach the breaking point, and strata fall into an altered equilibrium, earthquake is simply the collective name of all the cracks and shakings and disturbances that happen. They are the earthquakes. But for me the earthquake was the cause of the disturbances, and the perception of it as a living agent was irresistible" (James, 1906). In the coming paragraphs, and following William James' definition, I reflect on the specific assemblages of knowledge practices, tools, and institutional settings that are mobilized in order to define some of the many earthquake existences.

When James wrote about experience of the 1906 earthquake, seismology was a very different field than it is today. In seismology, discoveries are driven by the observation of large earthquakes and the history of the discipline has been a history of co-construction of the earthquake as scientific object. Some of the earliest-known scientific comments regarding earthquakes occurred in the mid-1600s, but most historiography on seismology starts soon after the Lisbon Earthquake in 1755 (Coen, 2013; Quenet, 2005). In Northern California, the first earthquake known as "the Big One" happened in 1868 in the still very rural Bay Area. The earthquake produced little damage and was left relatively undocumented but, following the event, the first seismometers were installed at UC Berkeley. When instrumentation was scarce and theory still in formation, science relied mainly on the descriptions of trained observers who were not always scientists or experts. No detail was too small and no nuance in the experience of a felt earthquake too trivial to be left undocumented. To define earthquakes, seismologists and local earthquake observers used their own perceptions of the event—"How did the earthquake feel?", as well as their sense of observation—"What did it produce?" (Coen, 2013). Whether interpreting their own observations or relying on the traces of a seismograph,[10] seismologists and observers made connections

[10]Both the Lick Observatory and the Students' Observatory in Berkeley were equipped with two Ewing seismographs and one Gray-Milne seismograph.

in order to establish relations between experience and science. They learned how to "read" an earthquake, how to organize sensations, observations, and recollections of a particular seismic event into a coherent explanation for an earthquake (November, Penelas, & Viot, 2009).

Working along the lines of both emotion and science, taking the materiality of the soft spots seriously, the experts I have met during this research are "amateur" (Gomart & Hennion, 1999; Hennion, 2004, Hennion, 2007). This definition indicates a person that possesses certain technological or scientific literacy (Chavrolin, 2013; Flichy, 2008; Greenberg, 2010; Haring, 2006) but, most importantly, who pays attention to several modes of existence—or several ontologies—of the risk and shows a deep attachment to the idea of making the Bay Area resilient to the forthcoming manifestation of earthquakes.

Earthquakes are not the same "objects" for scientists today as they were for seismologists 50 years ago. They are also becoming a different entity when taking into consideration the potential damage of the earthquake to come, the fault mechanism that triggers the movement, the ensemble of calculations that have been made in relation to the probabilities noted earlier, the way people live with an active seismic fault, or the memory of past events. Earthquakes and the risk they represent are never completely predefined. Instead, they have a changing definition that is the result of their complex trajectory—from the moment of recognition to being identified, and then, potentially, to being mitigated. To capture what it means to be *Waiting for the Big One*, this work unfolds different moments of earthquake instauration, its multiple concomitant occurrences: a past or a future event, a distant or close phenomenon, an object of science, worries and jokes, a geophysical process or an unrealized prediction.

1.3.1 Defining Souriau's Concept of Instauration

Observing the moment when the urgency to apprehend is not so pressing, before the disaster happens, when it is still a risk, invisible but present, is a good starting point. I focus on the relations between human and non-human, but also between different "modes of existence;" time and space between earthquakes are populated by researches and policy-making endeavours, thoughts, feelings, remembrances, anticipations, and speculations.

Following the path opened up by James, Souriau worked on the definition of "instauration," exploring what he called the "in-between-ness" of things that bear account of the quality of relations between two actants and defines instauration. Souriau's concept allows for the coexistence of sociotechnical assemblages—along with other form of "existences." Instauration is an action of doubling: experts do not construct earthquake risk, neither does this risk produce experts; rather, it is the "oscillation" (Latour, 2013: 166) between them that brings all earthquake risk to existence. Therefore, experts and residents "instaure" the risk of earthquakes, which leads to a transformation of not only the territory, but also the categories of knowledge and expertise (Chateauraynaud, 2008)[11]: "Instauration, in this case, designates the experimental schemes, the active preparation of observation, the production of facts, endowed with the power of showing whether the form realized by the arrangement is able to catch it, or not" (my translation; Stengers & Latour, 2009: 17).[12] To this concept of instauration, I add the concepts of distance (distanciation) and space (spatialization) to make visible the transformations created by the various ways in which the earthquake is instaured. Creating a conceptual framework that connected key notions of geography (space, scale, territory, distance, but also dwelling and remembering), studies of risk (vulnerability, resilience, and hazards), and STS (translation, mediation, and networks), these researches traced the dynamic emergence—or disappearance—of risk.

Following threads of concepts developed by ANT, these modes of existence (or artefacts and actants), can be organized in the form of a network. Callon's concept of "network" has been described as a system of relations that connects problematic statements.[13] The network,

[11]As Latour explained, making science and making art are actually comparable activities: "What fascinates Souriau about art (and what fascinates me about the laboratory), is the doing of making [*le faire faire*], the making exists, or in other words the replication and redundancy. It is the artist (or researcher) bouncing off the action and the reception of the work (or the autonomy of the fact)" (Latour, 2011: 9).

[12]"*Instauration, dans ce cas, désigne les dispositifs expérimentaux, la préparation active de l'observation, la production de faits dotés du pouvoir de montrer si la forme réalisée par un dispositif est ou non apte à les saisir*" (Stengers & Latour, 2009: 17).

[13]Callon's interest was to describe the circulation of scientific knowledge in relation to cultural and economic practices (Brown, 2002; Callon & Law, 2005; Callon, Méadel, & Rabeharisoa, 2000; Callon, 1986, 1997).

considered as an operating category, allows artefacts and actants to circulate between micro and macro structures in which they can simultaneously be global or local.

Looking at the range of articulations possible between artefacts and actants, researchers use metaphors as Souriau to define "prepositions" or "translation,": "defining an equivalence between heterogeneous objectives within particular actors. Its function was to overpass the false alternative between internalism and externalism" (Dosse, 1995: 27). "Translation," for ANT researchers, describes interactions, the event of "communication-transformation" between actants connected to one another. This system was first elaborated by Michel Serres[14] and, along with the work by Michel Callon, later became the theory of "Network-Actors" before evolving more fully and taking form under the name of "Actor-Network Theory" in 1999 (Callon, 1986; Callon & Ferrary, 2006).

Within these relational system, democratic principles prevail: no object has, in its essence, more power than any other. In addition, in a counterpoint to Aristotelian philosophy, which attributes substance to some objects, specifically the natural ones (e.g., flowers, birds, or even people's souls), "Latour grants all actants an equal right to existence, regardless size or complexity, all natural and artificial things must count as actants as long as they have some sort of effects on others things" (Harman, Bryant, & Srnicek, 2009: 17). In this context, power does not exist by itself; it is a combination of actants that may or may not hold for the long term: "Because of this notion of network, we can see how a point, which was isolated, becomes a control point for a large number of other points, and become a place of power. We can both follow the composition of power and its decomposition" (Callon & Ferrary, 2006).

[14]Paul Ricoeur also worked on the question of translation. As Domenico Jervolino notes, Ricoeur's translation is linked to the possibility of difference. Jervolino makes an interesting reference to the myth of the Tower of Babel, long understood as a malediction. For more contemporary readers, the myth of the Tower of Babel is about the recognition of the necessity for a plurality of humanity: "Translation becomes the privileged moment of a reconstitution of the plural unity of human discourse, opening the path for an ethic of verbal hospitality and conviviality" (Jervollino, 2006). Jervolino clearly sees in Ricoeur's concept the potentiality of a political form of non-violence interaction. I shall discuss the political consequence of ANT later in this work.

1.3.2 About the Following Chapters

In the following chapters, I look at the complex system of relations that instaure the risk of earthquakes, unearthing the role of emotions and everyday practices in the making of the space, the definition of risk, the creation of a relevant science and its translation into public policies. Considering the rapid propagation of the concepts and methodologies of risk reduction and disaster mitigation, and their constant interactions with everyday life, I question what space, knowledge, and expertise are created through the circulation of information, discourse, and interactions of actors in multi-risk, or post-disaster, environments. Looking for the trajectory of earthquake risk requires to identify where it appeared or disappeared, why its presence is sometimes overwhelming and sometime forgotten, and how people have regarded it during the relatively short, but intense, histories of both seismology and urbanization in the Bay Area.

I have built the present work in a symmetrical way following Callon's principle of generalized symmetry: "the rule which we must respect is not to change registers when we move from the technical to the social aspects of the problem studied" (Callon, 1986). Following the grammar of the network described earlier, I start with the local spatial dimension of the risk, progressively moving toward the scientific construction of earthquake risk. Each chapter has been defined as a specific moment of instauration of earthquake risk leading to a (complete or incomplete) transformation its environment.

Following this line of questioning, Chapter 2, *Defining a Conceptual Framework*, focuses on the main theoretical resources that have been used in this exploration of the instauration of earthquake risk. Examining the ways in which risk and disaster have been defined over the last several decades to articulate scientific knowledge and living experience, I analyse how researchers have tied together concepts of risk and disaster, as well as the question of individual and collective responses to threats. Then, I focus on questions of definition is a constant preoccupation of researchers, who are continually trying to refine the scope of this particular object of research and its complexes, changing relationship with other artefacts and actants. In an attempt to follow William James's steps, I also question how the different dimensions of earthquakes, from scientific definitions to risk evaluations, can be articulated regarding its specific experiences. From this perspective,

I look at how ANT provides the necessary theoretical backbone to describe what a future Big One could look like. Next, I argue that certain approaches in human geography and ANT can help elaborating our concepts, providing a perspective of situated knowledge, and giving an active role to the question of distance. Finally, we'll see that these perspectives can open up interesting discussions regarding the question of expertise.

Considering that earthquake risk is not always a tangible presence, Chapter 3, *The Elusive Dimensions of Risk and Disaster in the Bay Area*, questions its visibility in the everyday environment. Since its urban beginnings, the Bay Area has been tied up in a utilitarian relationship with the environment, both denying the materiality of the space and exploiting its resources, making earthquake risk difficult to "read." Looking for traces of earthquakes in the landscape, and through memories of long-time residents and commemorations, I argue that, even if earthquakes had had a strong influence on shaping the space of the San Francisco Bay Area, their existence can be hard to acknowledge and recognize. I also explore how "footprints" of earthquake risk emerge from traces left by previous disasters. A discussion of the Oakland Fire will be used as an example of an unfolding natural disaster and the methods of reconstruction. I then consider how controversies about scale, conversations about the legacies of past disasters, urban practices inherited from the time of the region's early urban planning, and relationships between residents and their territory add another layer of entanglement to the definition of earthquake risk.

Next, I focus on the experience of living with earthquake risk, and its capacity to transform the categories of knowledge, experience, and subjectivity. In Chapter 4, *Living with Risks*, I argue that the quality of the idea of an earthquake, as a potential concern or a threat, can only be found while looking at the density of experiences in the everyday coexistence with active seismic faults. In this chapter, I show that inhabitants and experts have instaured earthquake risk according to different modes of existence (Latour, 2011, 2013; Souriau, 2009), inviting us to take into consideration not only the multiple—and sometime contradictory—forms of risk experience in the hybridization of scientific knowledge, but also those that can transform the contour of a subject "at risk." In this chapter, I focus on the operations of translation and mediations through which, in the absence of recent major earthquakes in the Bay Area, earthquake still maintain their presence. I also explore how emotions and concerns about distant earthquakes (e.g., Christchurch, Haiti, Tōhoku) or smaller ones have contributed to the progressive instauration of local risk. Looking at the hybridization of knowledge between science and experience, this chapter focuses on the transformative aspects of earthquake risk for experts and scientists and their becoming into Earthquake Junkies.

Chapter 5, *The Case for Not Letting San Francisco Collapse*, discusses the work of these Earthquake Junkies over the last several decades. Taking a view contrary to that of Mike Davis and his famous text "The Case for Letting Malibu Burn" (Davis, 1998), I explore how experts and scientists have joined forces hoping they can mitigate the consequences of a future, large-scale earthquake. I discuss the fact that the possibilities for action, and for improvement of actual mitigation projects, lies in the articulation of different existences of an earthquake for different publics. Focusing this time, on the anthropology of expert elaboration on earthquakes and risk as an object of science, and its translation into public policy and scientific experimentation, I demonstrate how decades of work by concerned citizens, scientists, and experts have despite sometimes mixed results, laid the foundation for risk awareness in this seismically volatile region.

Finally, Chapter 6 explores the slow emergence of sociotechnical infrastructures designed to improve earthquake risk resilience in the San Francisco Bay Area. Building on in-depth interviews with experts, and following their professional and personal trajectories, the chapter describes resilience as a syncretic social process: the potential result of years of hard work and risk mitigation policies. Following a historical perspective and building on some important moments of earthquake risk prevention in the Bay Area of San Francisco, the chapter argues that resilience emerges from the collective development of a particular form of attention to the risk. In this context, it is the expert's experiences of living with the earthquake—of waiting for it, fearing it, remembering it, and getting ready for the next one—that gives sense to these complex sets of actions, defining the contours of a collective space of risk and allowing for resilient sociotechnical infrastructure to emerge or, if forgotten, to collapse.

Notes

i. That day, the panel discussion included, among others, Peter Hayes, Nautilus Institute, USF; Joonhong Ahn, Nuclear Engineering, UC Berkeley; Mary Comerio, Architecture, UC Berkeley; Cathryn Carson, Professor, History and then—Associate Dean of Social Sciences, UC Berkeley.

ii. While experts in conference rooms were pleading for democracy, the case for science was far from being settled in other public spaces. Debates raged about the scientific assertions and assumptions of the possible impact of radioactivity on the Bay Area. Pieces of information were collected by residents from around the globe. Social media provided a platform for the dissemination of alternative information and independent research (Abe, 2013) at a time when there was a significant absence of official information (Slater et al., 2012).

iii. This panel had been put together with the invaluable support of Prof. Joonhong Ahn (1958–2016), member of the Nuclear Engineering Department and Faculty Member of the Center for Japanese Studies within the Institute of East Asian Studies. Professor Ahn's dedication to the question of resilience and to sciences and technology studies (STS) approaches made his contribution to the field unique.

References

Abe, Y. (2013). *Why safecast matters: A case study in collective risk assessment.* A STS forum on Fukushima. Berkeley. Retrieved from http://fukushimaforum.wordpress.com.

Ahn, J., Carson, C., Jensen, M., Juraku, K., Mizokami, S., & Kumagai, Y. (2015). *Reflections on the Fukushima Daiichi nuclear accident.* Heidelberg and New York: Springer. https://doi.org/10.1007/978-3-319-12090-4.

Ahn, J., Guarnieri, F., & Furuta, K. (2017). *Resilience: A new paradigm of nuclear safety: From accident mitigation to resilient society facing extreme situations.* Heidelberg and New York: Springer. https://doi.org/10.1007/978-3-319-58768-4.

Akera, A. (2007). Constructing a representation for an ecology of knowledge: Methodological advances in the integration of knowledge and its various contexts. *Social Studies of Science, 37*(3), 413–441. https://doi.org/10.1177/0306312706070742.

Amir, S. (2018). *The sociotechnical constitution of resilience: A new perspective on governing risk and disaster.* London and New York: Palgrave Macmillan.

Anderson, B., & Harrison, P. (2010). *Taking-place: Non-representational theory and geography.* Surrey: Ashgate.

Bossu, R., Gilles, S., Mazet-Roux, G., Roussel, F., Frobert, L., & Kamb, L. (2011). Flash sourcing, or rapid detection and characterization of earthquake effects through website traffic analysis. *Annals of Geophysics, 54*(6), 716–727. https://doi.org/10.4401/ag-5265.

Bondi, L. (2005). Making connections and thinking through emotions: Between geography and psychotherapy. *Transactions of the Institute of British Geographers, 30*(4), 433–448. https://doi.org/10.1111/j.1475-5661.2005.00183.x.

Boudia, S., & Jas, N. (2007). Risk and risk society in historical perspective. *History and Technology, 4,* 317–331.

Brechin, G. (2006). *Imperial San Francisco, urban power, earthly ruin.* Berkeley: University of California Press.

Breviglieri, M. (2006). Penser l'habiter, estimer l'habitabilité. *Tracés, 23,* 9–14.

Brown, S. (2002). Michel Serres: Science, translation and the logic of the parasite. *Theory, Culture & Society, 19*(3), 1–27.

Callon, M. (1986). Some elements of a sociology of translation: Domestication of the scallops and the fishermen of Saint-Brieuc Bay. In J. Law (Ed.), *Power, action and belief: A new sociology of knowledge?* (pp. 196–223). London: Routledge.

Callon, M. (1997). *Actor-network theory—The market test.* Actor Network and After' Workshop.

Callon, M., & Ferrary, M. (2006). Les réseaux sociaux à l'aune de la théorie de l'acteur-réseau. *Sociologies Pratiques, 2*(13), 37–44. https://doi.org/10.3917/sopr.013.0037.

Callon, M., & Law, J. (2005). On qualculation, agency, and otherness. *Environment and Planning D: Society and Space, 23*(5), 717–733. https://doi.org/10.1068/d343t.

Callon, M., Méadel, C., & Rabeharisoa, V. (2000). L'économie des qualités. *Politix, 13*(52), 211–239. https://doi.org/10.3406/polix.2000.1126.

Chateauraynaud, F. (2008). Les mobiles de l'expertise Entretien avec Francis Chateauraynaud. *Experts, 78,* 4.

Chavrolin, F. (2013). Pense-bêtes, astuces et recettes de jardiniers-observateurs de papillons. Retour sur une science citoyenne. *Revue d'Anthropologie Des Connaissances, 7*(2), 485–500.

Coen, D. (2013). *The Earthquake observers: Disaster science from Lisbon to Richter.* Chicago: The University of Chicago Press.

Datson, L., & Galison, P. (2010). *Objectivity.* Cambridge: Zone Books.

Davis, M. (1998). *Ecology of fear: Los Angeles and the imagination of disasters.* New York: Vintage Books.

Davidson, J., Bondi, L., & Smith, M. (2005). *Emotional geographies.* Aldershot: Ashgate.

de Certeau, M. (1988). *The practice of everyday life*. Berkeley: University of California Press.

Dosse, F. (1995). *L'Empire du Sens. L'Humanisation des Sciences Humaines*. Paris: La Decouverte.

Flichy, P. (2008). *The internet imaginaire*. Cambridge: MIT Press.

Frankel, H. (2012). *The continental drift controversy: Evolution into plate tectonics* (Vol. 4). Cambridge: Cambridge University Press.

Fressoz, J.-B. (2008). Comment sommes nous devenus modernes? Petite histoire philosophique du risque et de l'expertise à propos de l'innoculation de la vaccine, 1750–1800. In S. Topçu, S. Cuny, & K. Serrano-Velarde (Eds.), *Savoirs en débat: Perspectives Franco-Allemandes* (Vol. 39, pp. 1–24). Paris: L'Harmattan.

Frickel, S., & Bess, V. (2007). Katrina, contamination, and the unintended organization of ignorance. *Technology in Society, 29*, 181–188.

Geschwind, C.-H. (2001). *California earthquake, science, risk and the politic of hazard mitigation*. Baltimore and London: The Johns Hopkins University Press.

Gomart, E., & Hennion, A. (1999). A sociology of attachment: Music amateurs, drug users. *The Editorial Board of the Sociological Review*. https://doi.org/10.1111/j.1467-954X.1999.tb03490.x.

Greenberg, J. (2010). *From Betamax to blockbuster: Video stores and the invention of movies on video*. Cambridge: MIT Press.

Guarnieri, F., & Travadel, S. (2018). *Un récit de Fukushima. Le directeur parle*. Paris: Presses Universitaires de France.

Haring, K. (2006). *Ham radio's technical culture*. Cambridge: The MIT Press.

Harman, G., Bryant, L., & Srnicek, N. (Eds.). (2009). *The speculative turn continental materialism and realism* (Anamnesis, pp. 304–334). Melbourne: re.press. Retrieved from http://books.google.com/books?id=coBDqJQeAQYC&pg=PA304&lpg=PA304&dq=Latour+Souriau&source=bl&ots=wHqRXjOttu&sig=GjPBBQjrUsrgrRgvWeykV9m6Q7Q&hl=en&sa=X&ei=XDFnUe_ZNoXmiwK7toDADQ&ved=0CGIQ6AEwAw#v=onepage&q=LatourSouriau&f=false.

Hasegawa, R. (2013). *Disaster evacuation from Japan's 2011 tsunami disaster and the Fukushima nuclear accident—Studies no. 05/13*. Paris. Retrieved from https://www.iddri.org/sites/default/files/import/publications/study0513_rh_devast-report.pdf.

Hennion, A. (2004). Une sociologie des attachements. D'une sociologie de la culture à une pragmatique de l'amateur. *Sociétés. Pratiques Musicales, 3*(85), 9–24.

Hennion, A. (2007). Those things that hold us together: Taste and sociology. *Cultural Sociology, 1*(1), 97–114. https://doi.org/10.1177/1749975507073923.

Hoffman, S. M., & Oliver-Smith, A. (2002). *Catastrophe and culture: The anthropology of disasters* (S. M. Hoffman & A. Oliver-Smith, Eds.). School of American Research Press and James Currey.

Houdart, S. (2008). *La cour des miracles: Ethnologie d'un laboratoire japonais* (Sociologie). Paris: CNRS Editions.

Houdart, S., & Thiery, O. (2011). *Humains non humains. Comment repeupler les sciences sociales?* Paris: La Découverte.

James, W. (1906). On some mental effects of the earthquake. *The Youth's Companion*. Reprinted in James, H. Jr, 1911, *Memories and studies* (H. James, Jr.). New York: Longmans, Green, & Co. (Vol. June).

Jervollino, D. (2006). *The future of phenomenology: Towards a philosophy of translation inspired by a phenomenological hermeneutics.* Retrieved from http://www.newschool.edu/nssr/husserl/Future/PartOne/Jervolino.html.

Johnson, L. (2011). *Insuring climate change: Science, fear, and value in reinsurance markets.* Berkeley: University of California.

Juraku, K. (2015). Why is it so difficult to learn from accidents? In *International Workshop on Nuclear Safety: From Accident Mitigation to Resilient Society Facing Extreme Situations.* Berkeley.

Kalmbach, K. (2015). From Chernobyl to Fukushima: The impact of the accidents on the French nuclear discourse. In T. Bohn & et.al (Eds.), *The Impact of disaster: Social and cultural approaches to Fukushima and Chernobyl* (pp. 67–96). Berlin: EB Verlag. Retrieved from https://research.tue.nl/en/publications/from-chernobyl-to-fukushima-the-impact-of-the-accidents-on-the-fr.

Knowles, S. (2011). *The disaster experts: Mastering risk in modern America.* Philadelphia: The University of Pennsylvania Press.

Knorr-Cetina, K. (1999). *Epistemic cultures: How the sciences make knowledge.* Cambridge, MA: Harvard University Press. Retrieved from https://books.google.com/books?id=WFEeib0Q9L0C&pgis=1.

Lamy, A. (2007). Récits médiatiques, mémoires électives. *EspacesTemps.Net.* Retrieved from http://espacestemps.net/document3673.html.

Lane, S., Odoni, N., Landstrom, C., Whatmore, S., Ward, N., & Bardley, S. (2010). Doing flood risk science differently: An experiment in radical scientific method. *Transaction of the Institute of British Geographers, 36,* 15–36.

Lash, S., Szerszynski, B., & Wynne, B. (1996). *Risk environment and modernity: Toward a new ecology* (L, Ed.) (Published). London: Sage.

Latour, B. (1991). *Nous n'avons jamais été modernes. Essai d'anthropologie symétrique* (Poche). Paris: La Découverte.

Latour, B. (2011). Reflections on Etienne Souriau's Les differents modes d'existence. In G. Harman, L. Bryant, & N. Srnicek (Eds.), *The speculative turn continental materialism and realism* (Anamnesis, pp. 304–334). Melbourne: re.press. Retrieved from http://books.google.com/books?id=coBDqJQeAQYC&pg=PA304&lpg=PA304&dq=Latour+Souriau&source=bl&ots=wHqRXjOttu&sig=GjPBBQjrUsrgrRgvWeykV9m-6Q7Q&hl=en&sa=X&ei=XDFnUe_ZNoXmiwK7toDADQ&ved=0CGIQ6AEwAw#v=onepage&q=LatourSouriau&f=false.

Latour, B. (2013). *An inquiry into modes of existence: An anthropology of the moderns*. Cambridge, MA: Harvard University Press.

Law, J., & Lien, M. E. (2013). Slippery: Field notes in empirical ontology. *Social Studies of Science, 43*(3), 363–378. https://doi.org/10.1177/0306312712456947.

Lazzarotti, O. (2006). *Habiter, la condition géographique* (Mapemonde). Paris: Belin.

Lévy, J., & Lussault, M. (2000). *Logique de l'espace, esprit des lieux. Géographie à Cerisy*. Paris: Belin.

Lewis, J. (2008, October 21). *The coming quake: Is Los Angeles ready for the Big One?* High Country News. Retrieved from https://www.hcn.org/issues/40.19/the-coming-quake/?b_start:int=1.

Lussault, M. (2007). *L'homme spatial, la construction sociale de l'espace humain*. Paris: Editions du Seuil.

Lynch, M. (2013). Ontography: Investigating the production of things, deflating ontology. *Social Studies of Science, 43*(3), 444–462. https://doi.org/10.1177/0306312713475925.

Mahony, M., & Hulme, M. (2016). Epistemic geographies of climate change. *Progress in Human Geography*. https://doi.org/10.1177/0309132516681485.

Marres, N. (2013). Why political ontology must be experimentalized: On eco-show homes as devices of participation. *Social Studies of Science, 43*(3), 417–443. https://doi.org/10.1177/0306312712475255.

Mialet, H. (2012). *Hawking incorporated: Stephen Hawking and the anthropology of the knowing subject*. Chicago and London: University of Chicago Press.

Mol, A. (2013). Mind your plate! The ontonorms of Dutch dieting. *Social Studies of Science, 43*(3), 379–396. https://doi.org/10.1177/0306312712456948.

November, V., Camacho-Hübner, E., & Latour, B. (2010). Entering a risky territory: Space in the age of digital navigation. *Environment and Planning D: Society and Space, 28*(4), 581–599. https://doi.org/10.1068/d10409.

November, V., & Leanza, Y. (2015). *Risk, disaster and crisis reduction*. Cham: Springer. https://doi.org/10.1007/978-3-319-08542-5.

November, V., Penelas, M., & Viot, P. (2009). When flood risk transforms a territory: The Lully effect. *Geography, 94*, 189–197.
Oliver-Smith, A., & Hoffman, S. M. (1999). In A. Oliver-Smith & S. M. Hoffman (Eds.), *The angry earth: Disaster in anthropological perspective* (1st ed.). Oxford, UK: Routledge.
Paquot, T., Lussalut, M., & Younes, C. (2007). *Habiter, le propre de l'humain. Villes, territoires et philosophie* (Armillaire). Paris: La Decouverte.
Perkins, J. B., Chakos, A., Olson, R. A., Tobin, L. T., & Turner, F. (2006). A retrospective on the 1906 earthquake's impact on Bay Area and California public policy. *Earthquake Spectra, 22*(S2), S237. https://doi.org/10.1193/1.2181527.
Quarantelli, E. L. (1998). *What is a disaster?: A dozen perspectives on the question.* London: Routledge.
Quenet, G. (2005). *Les Tremblements de terre aux dix-septième et dix-huitième siècles. La naissance d'un risque.* Seyssel: Champ Vallon.
Shineha, R., & Tanaka, M. (2017). Deprivation of media attention by Fukushima Daiichi nuclear accident: Comparison between national and local newspapers. In J. Ahn, F. Guarnieri, & K. Furuta (Eds.), *Resilience: A new paradigm of nuclear safety: From accident mitigation to resilient society facing extreme situations* (pp. 111–125). Heidelberg and New York: Springer. https://doi.org/10.1007/978-3-319-58768-4.
Slater, D., Keiko, N., & Kindstrand, L. (2012). Social media, information and political activism in Japan's 3.11 crisis. *The Asia-Pacific Journal: Japan Focus, 10*(24). Retrieved August 12, 2019, from https://apjjf.org/2012/10/24/David-H.-Slater/3762/article.html.
Solnit, R. (2009). *A paradise built in hell: The extraordinary communities that arise in disaster.* London: Viking.
Souriau, E. (2009). *Les différents modes d'existence.* Paris: Presses Universitaires de France.
Stengers, I., & Latour, B. (2009). Le sphinx de l'oeuvre. In *Les différents modes d'existence: Suivi de Du mode d'existence de l'oeuvre à faire* (pp. 1–75). Paris: Presses Universitaires de France.
Stock, M. (2004). *L'habiter comme pratique des lieux géographiques.* Retrieved from http://www.espacestemps.net/articles/lrsquohabiter-comme-pratique-des-lieux-geographiques/.
Strong, T. (2008). A review of Lorraine Daston and Peter Galison's objectivity. *The Weekly Qualitative Report, 1*(10), 62–66.
Taira, T., Silver, P. G., Niu, F., & Nadeau, R. M. (2009). Remote triggering of fault-strength changes on the San Andreas fault at Parkfield. *Nature, 461*(7264), 636–639. https://doi.org/10.1038/nature08395.
Tierney, K. J. (2001). *How will social science help us deal with earthquake?.* Chicago: University of Chicago Press.

Thien, D. (2005). After or beyond feeling? A consideration of affect and emotion in geography. *Area, 37*(4), 450–454. https://doi.org/10.1111/j.1475-4762.2005.00643a.x.

Tobriner, S. (2006). *Bracing for disaster: Earthquake-resistant architecture and engineering in San Francisco, 1838–1933*. Berkeley: The Bancroft Library, University of California Berkeley.

Vincent, C. (2017, December 29). En 2018, vous ne verrez plus le réchauffement climatique comme une fatalité grâce à la philosophe Emilie Hache. *Le Monde*. Retrieved from http://www.lemonde.fr/idees/article/2017/12/29/en-2018-vous-ne-verrez-plus-le-rechauffement-climatique-comme-une-fatalite-grace-a-la-philosophe-emilie-hache_5235552_3232.html#vZtFGWSXHFPtoISU.99.

Von Winterfieldt, D., Roselund, N., & Kisuse, A. (2000). *Framing earthquake retrofitting decisions : The case of hillside home in Los Angeles*. Pacific Earthquake Engineering Research Center. Retrieved from http://peer.berkeley.edu/publications/peer_reports/reports_2000/0003.pdf.

Walker, R. (1990). The playground for US capitalism? The political economy of San Francisco Bay Area in the 80's. In M. Davis, S. Hiatt, M. Kennedy, S. Ruddick, & M. Sprinker (Eds.), *Fire in hearth: The radical politics of place in America*. London: Verso.

White, G. (1945). *Human adjustment to floods—A geographical approach to the flood problem in the United-States*. The University of Chicago. Retrieved from http://www.colorado.edu/hazards/gfw/images/Human_Adj_Floods.pdf.

Wisner, B., Blaikie, P., Cannon, T., & Davis, I. (2006). *At risk*. London: Routledge.

CHAPTER 2

The Multiple Existences of Earthquake Risk

2.1 Thinking with Disaster

Attention to examining the multiple and what seem to be increasingly complex impacts of disasters and the challenges they make visible in large metropolitan areas is not recent. One of the most famous historical examples is the 1755 Lisbon earthquake, which had a considerable influence on the emergence of scientific knowledge. It transformed the social, philosophical, and metaphysical paradigms of the time—marking, for some, the beginning of the European Enlightenment (Amador, 2004; Coen, 2013; Dynes, 1997; Favier & Granet-Abisset, 2009; Fressoz, 2012; Quenet, 2005; Walter, 2008). Gilles Deleuze recalled the importance of the event which, had an impact on intellectual and metaphysical framing equivalent to the Holocaust in the twentieth century:

> It is very curious that in the eighteenth century, it is the Lisbon earthquake which assumes something like that, when across Europe people said: how is it still possible to maintain a certain optimism founded on God? You see, after Auschwitz raised the question: how it is possible to maintain a fading optimism about human reason. After the Lisbon earthquake, how is it possible to maintain the fading belief of rationality in divine origin?[1] (My translation, Deleuze, 1987)

[1] "*Il est très curieux que au dix-huitième siècle, ce soit le tremblement de terre de Lisbonne qui assume quelque chose de cela, où toute l'Europe s'est dite: comment est-il encore possible de maintenir un certain optimisme fondé sur Dieu. Vous voyez, après Auschwitz retentit la*

C. Mazel-Cabasse, *Waiting for the Big One*,
https://doi.org/10.1007/978-3-030-15289-5_2

In the twenty-first century, great earthquakes continue to provoke shifts, forcing us to face the limited human capacity to deal with and respond to such events. As French geographer Michel Lussault observed during the aftermath of the 2004 Indian Ocean tsunami: when engaged with destructive events, thinking becomes complicated by the influx of overwhelming information that seems incompatible with science's need for impartiality and objectivity:

> Soon, emotion facing what appeared as incomparable tragedies became global. [...] The extension at the scale of the globe of the rumors of the disaster came together with the diffusion of a spectacular dramaturgy associating narratives, pictures of professionals and amateurs, more or less scientific description of the tsunami and its consequences, anticipation discourses on the conceivable future replicas of such phenomena.[2] (My translation, Lussault, 2007: 15)

Since the Indian Ocean earthquake and tsunami, other major disasters, including a great number of earthquakes, have not only deeply transformed the actual space that they impacted, but also the world around them. Among others, they include Hurricane Katrina in 2005, the Haiti earthquake in 2008, and, in 2011, the Christchurch earthquake, and the Tōhoku earthquake, tsunami, and nuclear disaster, to name just a few.

These events have raised concerns about the capacity for individuals and collectivity to face the disaster, to cope with the aftermath, and to remain resilient. They have also shown that disasters do not respect borders; they are not limited to one space, one political context, or, for the researcher, one academic discipline. Because of their incredible power, they can also change larger understandings of epistemologies and metaphysics, as they have done in the past. Each time they occur,

question: comment est-il possible de maintenir le moindre optimisme sur ce qu'est la raison humaine. Après le tremblement de terre de Lisbonne, comment est-il possible de maintenir la moindre croyance en une rationalité d'origine divine?" (Deleuze, 1987).

[2] "*Rapidement, l'émotion face à ce qui apparut comme une tragédie incomparable devint mondiale [...] La dilatation à l'échelle du globe et l'écho de la catastrophe s'accompagna de la diffusion d'une dramaturgie spectaculaire, associant des récits, des images de professionnels et d'amateurs, des descriptions plus ou moins scientifiques du tsunami et de ses conséquences, des discours d'anticipation sur les futures répliques envisageables d'un tel phénomène*" (Lussault, 2007: 15).

disasters convene a kaleidoscope of both micro- and macro-events that engage both human and non-human agents at different scales and times. After each occurrence, the shock provoked by their violence—including the tragedies of lives disrupted or ended; the financial cost of the destruction; and the amount of time, energy, and money needed to return to "normal"—surpasses the predictions and, often, even the imaginations of victims, observers, and researchers. More often than not, the consequences of such large events last for decades—not only visible in the urban landscape, but also alive in people's memories. Each disaster leaves its own trace but, each time, the tribute paid is the same: destruction, loss, and grief.

This "sudden emergence of environmental issues on the political agenda [... and] the multiplication of tangled objects, which can no longer be confined to the natural world or naturalized by anyone" (Latour, 2004, as quoted by November, Penelas, & Viot, 2009: 195) has pushed researchers to redress the question raised by risk and disaster (Collectif, 2008a, b), reconnecting, for some, with the field of early disaster studies (Quarantelli, 1999; Quarantelli & Perry, 2005; Tierney, 1994). Historians have long shown how the Industrial Revolution created not only the conditions for the creation of innovative technologies designed to reduce risk, but also the narratives that denied the materiality of these conditions to support the idea of "progress" (Favier & Granet-Abisset, 2009; Fressoz, 2012).[3] In so doing, industrialization, urbanization, and modern life have created the conditions of their own vulnerabilities (Beck, 2009; Lagadec, 1987).

In recent years experts and lay people alike have questioned the horizon of a world without disaster, as well as the place that disasters take in our lives (Dupuy, 2002; Latour, 2013; Neyrat, 2008; Serres, 2009; Stengers, 2009). Acknowledging that the balance between dangers and safety precautions is constantly renegotiated in contemporary societies (Lupton, 1999a, b; Wiener, 2010; Wiener & Rogers, 2002), the role

[3]Douglas pointed out: "Risk has largely replaced older ideas about the cause of the misfortune. Concepts such as sin, which were once used to provide explanation for misfortune, are now discredited. In their place, is the 'modern, sanitized discourse of risk' which 'is all we have for making bridges between the known facts of the existence and the construction of a moral community [...] Indeed, risk provides secular terms for rewriting scripture: not the sins of the fathers, but the risks unleashed by the fathers are visited on the heads of their children, even on the nth generation" (Douglas 1992: 26; as quoted by Lupton, 1999b: 47).

of expertise in the definition of such concepts has been as well underlined (Callon, Lascoumes, & Barthe, 2009; Haas, 1992). Interrogations related to the threat of potential disasters—coming together under the concept of "risk"—have become an extensive field of research where the seminal work of Ulrich Beck (1992) has been both celebrated and criticized (Bourg, Joly, & Kaufmann, 2013; Lash, Szerszynski, & Wynne, 1996; Latour, 2004).

Building on these perspectives, I use a definition of risk that takes into consideration the semantic and operative dimensions of the concept risk. Five of these seem particularly important, since they are widely used across disciplines. The *Stanford Encyclopedia of Philosophy* provides the following definition:

1. Risk: an unwanted event which may or may not occur.
2. Risk: the cause of an unwanted event which may or may not occur.
3. Risk: the probability of an unwanted event which may or may not occur.
4. Risk: the statistical expectation value of an unwanted event which may or may not occur.
5. Risk: the fact that a decision is made under conditions of known probabilities ("decision under risk" as opposed to "decision under uncertainty") (Hansson, 2012).

Moving from the threat to the actual disaster, anthropologists have also looked at past events (Clavandier, 2004; Henry, 2005; Hoffman, 1994; Hoffman & Oliver-Smith, 2002; Langumier, 2008; Revet, 2007) and everyday risks (Tulloch & Lupton, 2003). From the perspective of the social sciences, some of the most important trends have questioned the intersection between risk, regulation, and governance (Borraz, 2008; Cabane, 2012; Jasanoff, 1986, 2010); technological responsibility (Clark 2006; Demuth, 2002; Jonas, 1995); and questions related to alerts, vulnerabilities, and the environment (Burton, Kates, & Whites, 1993; Charvolin, 1997; Creton-Cazanave, 2010; D'Ercole, Gluski, Hardy, & Sierra, 2009; Gralepois, 2008; Hoffman, 1998; Miller et al., 2010; Pelling, 2003; Reghezza-Zitt, 2013; Watts & Bohle, 1993). Adding to the previous definition, researchers today generally acknowledge that "disasters come into existence in both the material and the social world and perhaps, in some hybrid space between them" (Oliver-Smith & Hoffman, 1999: 24).

2.1.1 The Apparent Paradox of the Bay Area

Talking about risk and disaster in the San Francisco Bay Area can sometimes feel a little inappropriate. "No, I don't think about that at all, I've lived in California my whole life, so earthquakes are not even part of my thinking" [M.3], said one of the respondents during our discussion, with the impatient look of someone waiting for his interlocutor to change the subject. Another interviewee, a survivor of the 1991 Oakland Fire, told me firmly: "I never thought that about the fire" [G.6]. The topic of our discussion almost felt like a discourteous opinion about an area that has so much else to offer: vibrant cities, picturesque history and political figures, Silicon Valley, spectacular landscapes, or even the temperate weather.

Back in 2008, freshly relocated to the Bay Area from Europe with a dissertation project about risk in hand, I discovered that what I thought was a problem—namely, the risk of earthquakes or fires in the Bay Area—was often considered an almost irrelevant subject by many of people whom I interviewed. "Most people I know don't think about it or, if they think about it and worry about it, I don't think it affects the way they live their lives," [R.4] said a third respondent, leaving me to reflect further on the subject.

Later in the interviewing process, short, "closed" sentences would come back often. When asked how they deal with the risk of earthquakes, common answers would be: "It is not something that I feel concern with" [G.6], or "That does not affect my day to day thinking at all" [M.3]. Connected to such statements, the idea prevails that residents of the Bay Area cultivate a certain detachment from earthquake risk. In fact, this distance between the possibility of a large-scale disaster and the residents' seeming disdain has been a recurrent complaint from risk specialists, who, in turn, have been quick to criticize San Francisco Bay Area residents for their general lack of interest in understanding their own risks. However, despite these apparently entrenched stances, recent research has shown that not only has the level of awareness among residents of the Bay Area tended to improve over time, but also that residents frame their reasoning into modeling, "probably reasonable,"[4] and verifiable matrices.

[4] "We identified a new set of problems with the earthquake safety of hillside homes; we know how to fix these problems, at least for moderate to strong earthquakes like the Northridge earthquake; as long as the costs of the fixes are significantly below the damage

For instance, following the Northridge earthquake, which hit the hillside homes of Los Angeles particularly hard in January 1994, researchers at the University of Southern California conducted a study on personal decisions frameworks regarding retrofitting (Von Winterfeldt et al., 2000). They defined types of decision-making framework that either coincided or conflicted with one another. The first is the "Regulatory Frame," in which the key question is "what caused the damage to hillside homes in past earthquakes (especially the Northridge earthquake) and what can be done to avoid similar damage in future earthquakes" (2000: 23). Using the responses to this question, the rule for action "appeared to be to spend no more money for retrofitting than what could be saved by avoiding damages in a major earthquake" (2000: 23). The second framework of analysis was called the "Homeowner Frame," which challenged the recommendations of the previous framework, raising the issues of costs and benefits of major retrofitting, and the low probability of a large-scale earthquake. They concluded that, "under three conditions, a minor retrofit becomes the preferred option: 1. If the time horizon is increased to about 30 years; 2. If the replacement value V is increased to about $750,000; 3. If the homeowner is very risk averse. No reasonable assumptions could make a major retrofit the preferred option" (2000: 29).[5]

In 2009, the Bay Area Chapter of the American Red Cross conducted a survey among 1201 residents San Francisco to evaluate their level of preparedness. The market-research company Issues & Answers was hired to conduct interviews in English, Spanish, Mandarin, and Cantonese. The results showed that people with previous experiences of a disaster (34% of the panel) were more likely to have taken one of several

that is likely to occur in a Northridge-like earthquake, and as long as these fixes provide a higher degree of assurance that there won't be any damage, they are probably reasonable" (VonWinterfeldt, Roselund, & Kitsuse, 2000: 18).

[5]Following this perspective, from the authors' vantage point, the more accurate solution in terms of public policy should be to implement policy that encourages homeowners to invest in minor earthquake retrofitting. "The primary reason for this switch is the longer time horizon. With a longer time horizon, the probabilities of moderate and strong ground shaking increase (approximately, but not precisely, by a factor of three), and the probability of low shaking decreases. As a result, retrofitting alternatives will become more attractive" (VonWinterfeldt et al. 2000: 31).

preparedness action steps as advocated by the American Red Cross (e.g., make a plan, build a kit, and get training). In this group, 89% stated that they were ready to face a possible disaster. Respondents with higher incomes ($50,000 and higher) and long-term residents (more than 10 years) were also more likely to take at least one preparedness action step. The research also showed that the level of preparedness was slowly increasing. In 2006, only 6% of respondents indicated that they had fulfilled the three steps suggested by the American Red Cross preparedness plan, whereas 22% had done so in 2009. In addition, 32% of the respondents had taken two steps and 29% only one. Only 16% of Bay Area residents took no safety measures in 2009, which is lower than the figure for 2008, which was 21%. The role of local and federal institutions was also well-known to respondents. Of the population with school-aged children, 75% knew about their children's school's emergency plan. The American Red Cross was trusted by 60% of respondents to be a good source of information, as well as a reliable actor in response to an emergency. In contrast, town and city mayors, the California state governor, and the Federal Emergency Management Agency (FEMA) were only trusted by approximately 33% of the respondents. Television networks were seen as the least accurate source of emergency information (Issues Answer, 2009).

The implication of such studies is important: first, because residents' decision-making frameworks are finally considered rational and, as such, can reach the public space; second, within a socially and scientifically acceptable perspective, they have put forward original, but also sometimes contradictory, positions, solutions, and recompositions of the space of risk and disaster in the Bay Area. These controversies, as we can call them, draw the (always temporary) lines between what matters and what does not; what should be taken into consideration and what need to be left aside.

2.1.2 *Situated Knowledge and Multiple Perspectives*

In the academic context, risk and disaster have been analyzed and studied by many disciplines that cross the fields of humanities, social, political and earth sciences. These studies have used multi-disciplinary methodologies taken from statistics, computer or engineering sciences, and anthropology.

This multiplicity of perspectives is further complicated by the fact that academic expertise is constantly in discussion with not only the more operational level of policy-making, but also with lay people, residents, consumers, citizens or concerned groups that are both the beneficiaries and the subjects of scientific knowledge. In addition, as was the case during the "STS Forum on the East Japan Disaster," each event brings together different levels of expertise and narration at that specific time and place.

This difficulty in deploying the complexity of each situation has, for several decades now, been one of the many research programs tackled by researchers in STS, which can be defined as an "interdisciplinary field that examines the creation, development, and consequences of science and technology in their cultural, historical, and social contexts" (Hackett, Amsterdamska, Lynch, & Wajcman, 2008, backcover). From this perspective, understanding how knowledge is defined and circulated has been one of the central questions of the field, building on what has also been developed previously by anthropology:

> A fundamental insight of anthropological studies of learning is that knowledge is not transmitted across generations as a ready-made corpus of information, but rather undergoes continual regeneration in the context of learners' practical engagement with their surroundings. Thus the contribution that each generation makes to the next lies in shaping the contexts, or providing the scaffolding, within which learners develop their own understanding. (Ingold, 2005)

This mobility of knowledge in the making of Science has been pointed by Hélène Mialet who recalled that Science, as a discipline but also as a narrative, has for long been "concealing" many aspect of its practices, in a effort to present an even – but unrealistic –appearances: "*the assistant are hidden (…), the conversation effaced (…), the memory is re-invented, the mythical account circulate (…)*" (Mialet, 2012: 4). In the continuity of these researches, the ideal of a scientist "detached" from his environment has been criticized by authors in who had favored the description of a hybrid methodologies, personae and objects (Haraway, 1988; Latour, 1991; Star & Griesemer, 1989) making visible situated knowledge made possible by a network of human and non-human (Houdart & Thiery, 2011) which encapsulate the description of the contexts and processes that make science possible (Houdart, 2008). Looking at the history of earthquake sciences, Deborah Coen recalls that, in the nineteenth century, the "scientific description of an earthquake was built of stories – stories from as many people, in as many different situations, as possible" (Coen, 2013: 3). To define earthquakes, seismologists used their own perceptions of the event ("How did the earthquake feel?"); and their sense of observation ("What did it produce?"). They also used as many indirect sources as they could, including accounts from the magazines and newspapers, testimonies of other observers, and, when available, measurements from scientific equipment.

In seismology, as in many other scientific domains, the progress of research in the twentieth century has favored the movement from subjective accounts to data centric approaches provided by measuring instrumentations, allowing the development of predictive models and the development of a probabilistic vision of earthquake risk. In the process, earthquakes have become more abstract objects of science, virtually modelised and defined by their quantifiable qualities. In such a context, it could have been expected that scientists and experts working on earthquake risk would grow more and more distant from the experience of residents (Bessy & Chateauraynaud, 1995; Mitchell, 2002). To borrow one of Ulrich Beck's book titles, in a *World at Risk* (see Beck, 2008), earthquakes could have finally come to be perceived like any other ungrounded, immaterial threat. Yet, knowledge about earthquakes has remained hybrid and grounded in experience, and this diversity of perspectives is precisely what makes it an object of scientific research.

2.1.3 Making the Risk Visible

Specialists of earthquakes who are also residents in this active seismic zone continue to frame earthquake risk as both an object of science and one of experience, pursuing a tradition of knowledge-making that is now considered "hybridized." For them, all sharing the same risk in the place where they live and work, earthquakes are, as they were for seismologists during the twentieth century, phenomena that "cannot be comprehended exactly" (Coen, 2013: 8), or need to be understood from different perspectives (James, 1906).

This hybridization of knowledge changes not only the definition of the science and knowledge, but also the subjectivity of experts, scientists, and residents. For instance, as I have discovered during my field research, some experts and scientists call themselves "Earthquake Junkies." This self-given nick-name reflecting on the almost addictive nature of their dedication of earthquake preparedness illustrating how they redefine a more engaged figure of expertise: a knowing subject, grounded in its environment, circulating between the layers of knowledge(s) that are usually proofed so as to distinguish the scientist, the expert, the lay person, or the amateur.

One of the instances of knowledge hybridization occurred in 2014, just after the South Napa earthquake,[6] when Joshua Bloom, a professor of astronomy at UC Berkeley, made the headlines of the local newspaper with an invention tucked into a sandwich box: a home-made, easy to build and cheap early warning system. "I thought it was silly that every time I closed my laptop, I couldn't get a warning" (Knobel, 2014), explained Bloom, who is an Early Warning Program Beta Tester, and who, through this invention, was reviving a *modus operandi* which was more usually associated with amateurs than university professors.

In a blog post reflecting on this invention, Bloom recalled that his device uses the latest development of the Early Warning Program that has been sending live alerts to test users since 2012: "In October 2013, I built a prototype earthquake early warning (EEW) device, for less than $110 in parts, for my house that taps into the California Shake Alert system. In the recent 6.0 magnitude quake centered near Napa, we got about 5 seconds warning here at my home in Berkeley, CA, before the shaking began" (Bloom, 2014). Asked what would be the next step for his invention, Bloom explained that he had placed the device in his kids' classroom and planned to improve his prototype, adding that his children find the project "super cool" (Knobel, 2014).

2.1.4 Waiting for the Big One

Waiting for the Big One is an active posture, encapsulating a large set of practices that articulate past, present, and future knowledge on different geographical scales: how memories of past disasters resonate with the risk of future ones, how expert knowledge emerges from past and distant

[6]The South Napa Valley earthquake was an M.6.0 event that hit the city of Napa, California, on August 24, 2014, at 3:20 a.m. local time.

experiences, and, finally, how the invisible threat transforms space, subjectivity, knowledge, and politics in and of the Bay Area. This event horizon—the next Big One—provides an important perspective on a set of operations that happen today, every day, in the Bay Area and which, from residential choices to seismic-mapping activities, from memorialization to projection, define and help bring the potential consequences of an earthquake into existence, into visibility.

Waiting for the Big One is also a collective experience that brings together both experts and non-experts, who collaborate to mobilize residents' attention and concern. Looking at the many different ways an earthquake exists—not only as a geological process, but also as a past and a potential future threat—help unfolding the everyday experiences of concerned individuals in the San Francisco Bay Area, and illustrate how a preconceived framework of analysis often fails to recognize the complexity of "living with risk." Following the many ways in which earthquake risk is approached by these Earthquake Junkies experts-residents, I consider the Big One as geographical and historical, collective and individual, scientifically and empirically made sociotechnical assemblage. Earthquake risk and the understanding of future earthquakes are the results of co-constructions that vary depending on the particular situation of the person or persons concerned.

What we have, following Souriau, called the instauration of the earthquake is a slow, but organised process that needs to be looked at within a proper historical perspective, while also considering the diversity of experiences. Looking for the traces of risk and disaster in the Bay Area, expounding on a complex map of emotions and memories, defines the dimensions and intensity of the San Francisco Bay Area earthquake risk.

This chapter examines some of the difficulties that have emerged when trying to define and localize risk and potential disaster in this region. Then, we'll look for the physical traces of earthquake risks.

This "on-site" research will allow us to identify some of the specificities of this risk, focusing on three materialisation of earthquake risk: first, the Bay Area fault maps, which center on the seismic faults; second, the California Memorial Stadium, which has crystallized some of the most entrenched controversies not only around the questions of risk and safety, but, additionally, around the ideas of conservation and heritage; and third, the San Francisco Bay Bridge. In this chapter, I have also shaped an ethnography of the Oakland Fire—an event of considerable importance regarding the understanding of regional disasters among the residents of the Bay Area.

2.2 The Denial of Topography and the Exploitation of Nature

In the short, but intense history of the urbanization of the San Francisco Bay Area, the sense of risk and the occasional disasters have been unwelcome guests. For 150 years, urbanization has paralleled economic development.[7] If different narratives have told the same story, they have—sometimes for opposing reasons—tended to hide the potentiality of natural disasters. One of these reasons could be that natural resources are what made California rich (Brechin, 2006; Gaar & Miller, 2006).

2.2.1 Nature as a Resource

A nostalgic lost paradise before its massive urbanization, San Francisco was an ecologically diverse ecosystem shaped by a delta, estuaries, wetlands, and strong tides. Fauna and flora were prolific and early accounts of the area described a "sky darkened by the flights of ducks and geese, salmon runs coming up the delta so thick that you had the impression you could walk across their backs all the way to the other shore" (Save San Francisco Bay, n.d.).[8]

Gold extraction in the relatively nearby hills was, of course, the fuel for the economic development of the Bay, but there was also systematic, organized exploitation of available natural resources including water (Reisner, 2003), wood and fish (Montgomery, 2004), and a variety of wild animals. Agriculture soon became a great resource for the state.

[7] "Resources extraction and economic development were reciprocal, the gains from resources extraction created prosperity within the region, and the regional special order was such as to maintain these conditions" (Walker & Thomas, 2010: 5).

[8] Most of the wildlife is, of course, no longer found along highways or in the backyards of residential areas: "tule elk, antelopes and grizzly bears, and thousands and thousands of migratory shorebirds – so many that duck hunters, up until the 1850s describe being able to kill twelve birds with one shot" (Save San Francisco Bay, n.d.). Wildlife is still present in the East Bay: mule deer, coyotes, gray foxes, raccoons, skunks, opossums, and fox squirrels live in the hills. Mountain lions are also part of the mix of the East Bay wildlands–urban interface, being regularly spotted in the Berkeley and El Cerrito hills.

The planting of melons and oranges was capable of generating agricultural empires and, with the help of cheap migrant labor, did so, in many cases. As a result, the Bay Area, as we know it today, is what is left (or what has grown) from the "voracious appetite" (Walker, 1998) for the resources of this promising "new frontier" land (Berglund, 2007). Natural resources were transformed into valuable commodities and, in the process, an industrial city grew from the ground. The transformation of the space was also a powerful fuel for innovation, and residents of the region have built up the area largely only thinking about its nature as a resource, and not as a threat.

Powered by this incredible growth, in the middle of the nineteenth century San Francisco was a city of 25,000 inhabitants, and boasted a large harbor and the capacity to supply trade to the rest of the continent. The process of planning the city's development was concomitant with the necessity to integrate San Francisco with America's larger project of "manifest destiny," transforming what was once a "frontier" into a model of civilized urban life. "Urbanity, not rusticity, was in fashion" (Walker, 1995). In a Haussmannian movement to give San Francisco the structure that the gold city was missing, middle- and upper-class houses were planned using a grid system, and were built with highly decorated facades reproducing a mix of European architectural styles. These Victorian homes were about to give San Francisco its distinctive characteristic.

In order to keep the pace of this growing urbanization, city planners made some trade-offs with nature. The first urban plan of San Francisco was structured on a rectangular grid and, in order to maintain its shape, successive planners had to treat the topography of the Peninsula with "cavalier disregard" (Tobriner, 2006: 9). Beds of freshwater areas were filled in, and then deleted from maps, while some of the hills and marshes were simply smoothed over.[9] This is how Yerba Buena Cove, once a tidal ecosystem, was filled in, and how 80-foot high sand hills were shoveled away. This is also how the Ricon Hills, known for their incomparable views, were leveled and the marshes under Mission Street,

[9] "According to one source, O'Farrell [the civil engineer in charge of the grid] at first blanked when asked to extend the grid without reference to topographical features, but in the end, he did. In accordance with standard practice in nineteenth century America, O'Farrell platted across tidelands that were regularly under water, over marshes, and up the slope of steep hills in order to facilitate land sale" (Tobriner, 2006: 9).

which were so unstable that animals would drown in them, filled into provide foundations for new buildings.[10]

With the real estate market being an important economic motor, topography could not be an obstacle.[11] Many of the water lots, sold by the municipal authorities which had been progressively filled in by developers and turned into "made ground," revealed themselves to be, unsafe in case of a earthquake. However, by the end of the nineteenth century, this "hard core" development was subject to contestations: new residents started to question the brutal ways of the city's planners.[12] They were not only concerned about the aesthetic quality of their living environment, but also by the matter of safety; Earthquakes had rattled the Bay Area in 1838 causing serious damage to the small adobe Mission built in the late eighteenth century. Later, in 1850, several earthquakes of small magnitudes[13] caused a great deal of damage in the "made ground" areas around the city, more so than in other parts.[14] Despite these concerns, weak building codes, mostly for residential homes, remained the norm, as developers were afraid of potential investors fleeing from too many constraints.

2.2.2 *The Multiple Ecologies of the Bay Area*

Today, the greater Bay Area remains one of the wealthiest areas of the whole country,[15] and exploitation of its natural resources has not

[10] Incredibly, in some places, between 40 and 80 feet deep.

[11] As Tobriner noted, the city was then selling lots in order to cover municipal expenses (Tobriner, 2006: 11).

[12] Mobilization by the citizens and the press had in little to no impact on the city plan, which was drawn according to the city's overriding business plan.

[13] The 1868 earthquake was an M6.8 earthquake, with its epicenter at the southern end of the Hayward Fault. The earthquake was, at that time, the most destructive in California's history. In San Francisco, five deaths and significant property loss was reported. A significant amount of damage also occurred in Alameda County, where a number of small rural cities were almost totally destroyed.

[14] After the 1856 earthquake in Alta California, an editorial stated, "Since the terrible shock of Friday last, speculation has been rife as to the probable effect it would have upon the minds of inhabitants and upon the property interests of the city" (Tobriner, 2006: 18).

[15] So-called "Silicon Valley," on the peninsula of the Bay Area, is geographically the third largest high technology center in the United States, behind New York City's greater

ceased[16]: large-scale civic mobilization has largely succeeded in successfully defending the remaining species and preventing wetlands[17] from being turned into oil terminals and freeways,[18] in the name of "development."[19]

Despite, or sometimes because of, these discords, San Francisco has been able to keep its singular character; unlike its southern sister, Los

metropolitan area and metropolitan area of Washington, DC. In the San Francisco Bay Area itself, it ranks first. The region is the most important producer of high technology devices in the United States. In 2009, the San Francisco Bay Area's GDP ranked it as the equivalent of the 22nd richest country in the world, had it been its own country.

[16]Since the first moment of its construction, San Francisco has been pulled apart by the exploitation of its resources on one side and the people's defense of the environment on the other. Bay Area conservation groups are numerous and very active. The Bay Institute of San Francisco, San Francisco Baykeeper, the California League of Conservation Voters, California Native Plant Society, Friends of the San Francisco Estuary, the Greenbelt Alliance, the Marine Mammal Center, and the Save the Bay organization, to name a few, have been very active in trying to protect the remaining space against the growing "concretization" of the greater Bay Area.

[17]Among others battles, opponents of non-regulated development of the Bay Area have been able to stop developers from filling in the Bay more fully. The first place to be filled in was the Yerba Buena area, around Mission Bay; soon, thereafter, every San Francisco creek and tideland. Then came the creation of Treasure Island, the Alameda Naval Air Station, Mills Fields, where the San Francisco Airport is located, and the San Pablo Bay fill project. By the 1960s, 91% of the San Francisco Bay wetlands were gone.

[18]With the freeway revolt—the first of its kind in the United States—neighborhood associations collected 30,000 signatures against the construction of the Embarcadero Freeway, which was meant to connect the Bay Bridge to the Golden Gate Bridge. By 1985, the removal of the freeway was supported by a large coalition of environmental organizations. The-mayor, Dianne Feinstein, supported the removal project. In 1991, the freeway was torn down and replaced by a boulevard with trolleys and a pedestrian promenade running along the waterfront. This area's real estate value grew by 300%. For more information about the removal of the freeway, preserving cities, and, specifically, the Embarcadero Freeway, see http://www.preservenet.com/freeways/FreewaysEmbarcadero.html.

[19]"Urban renewal legislation passed by Congress in 1949 and 1954 gave cities the power to assemble land, clear it of offending uses, and finance redevelopment. San Francisco, like all big cities established a Redevelopment Agency to spearhead its efforts. Justin Herman directed that agency aggressively for many years, backed by the San Francisco Planning and Urban Renewal Association [a citizen group], and later the Convention and Visitors Bureau [an arm of the hotel and tourism industry]" (Walker, 1998: 4). The agencies targeted low-income areas and, as a consequence, destroyed more than 10% of the Victorian houses that were to become city landmark buildings. During the 1950s and 1960s, 20,000 people were delocalized. Other development projects targeted those living in

Angeles, it has never embodied an "ecology of fear"[20] (Davis, 1998). For long time, the city of San Francisco has kept its character—as did Oakland and Berkeley, and to some extent, Albany, El Cerrito, and Richmond[21]—avoiding, although not completely, the aesthetically destructive effects of suburbanization during the second half of the twentieth century (Walker, 1998). Nonetheless, the landscape (Groth & Bressi, 1997) and the cityscape (King, 2011) today reflect the story of rapid urbanization, and its economic and social transformations—reminders of the diversity of everyday individual experiences, goal and practices.

Today metropolitan area is a confluence of millions of lives, each having its own map,[22] its own rhythm, and its own path. Putting all of the pieces of this urban puzzle together, all of these different worlds, is a difficult enterprise (Solnit, 2010; Walker, 1995). The ecology of the Bay

hotel housing, mostly occupied by single Filipino, White, and African-American men who had retired from working on the docks. This renewal marked the premise of the property boom, which lasted for a little more than a decade, between the 1960s and the first major oil crisis of 1973. Among the famous cases of preservation that were embraced and led by citizen-based organizations was the preservation of San Francisco's cable car, achieved by its receiving National Landmark Status; the "freeway revolt" in 1955, which stopped the freeway from running through Golden Gate Park; and the first loft emerged from Ghirardelli's former chocolate factory. In 1968, the African-American community living in the Fillmore District of San Francisco won its battle against an urban renewal project, allowing them the right to protect their homes from destruction, which was a historical first in the USA.

[20] On this question, see also *Insecurity and Segregation: Rejecting an Urbanism of Fear* (Pattaroni & Pedrazzini, 2010).

[21] This is probably more visible in Oakland and Berkeley than in Albany, El Cerrito, and Richmond. El Cerrito and Richmond, in particular, have seen larger suburban development, targeting both the lower- and upper-middle classes in these areas. Both cities, however, have succeeded in keeping some of their historic buildings.

[22] "San Francisco has eight hundred thousand inhabitants, more or less, and each of them possesses him or her own map of the place, a world of amities, amours, transits routes, resources and perils, radiating out from home. But even to say this is to vastly underestimate. San Francisco contains many more than eight hundred thousand living maps, because each of these citizens contains multiple maps: areas of knowledge, rumors, fears, friendship, remembered histories and facts, alternates versions, desires, the maps of everyday activities versus the map of occasional discovery, the past versus the present, the map of this place in relation to others that could be confined to a few neighborhoods or could include multiple continents of ancestral origin, immigration routes and lost homelands, social ties or cultural work" (Solnit, 2010: 3).

Area can be found in the interwoven stories, the multiple landscapes and landmarks, the living being, and the projects and ideas generated from the area.

It is a moving ecology, as Solnit defined it, of an *Infinite City* (Solnit, 2010), born from the cohabitation of atomic bombs and cypress trees, butterflies and long-closed jazz clubs, coffee places and murder scenes, murals and creeks, and, of course, the multitudes of "tribes" that inhabit, or have inhabited, the place: joggers, Chinese, the homeless, students, African-Americans, hipsters, Mexicans, civil servants, the Irish, or the city's gay community.

Solnit used crossed maps of San Francisco and the Bay Area to reflect the invisible strata of this urbanity. Her maps make the invisible visible—the remote presence of lost worlds, from the first homes of the Native American dwellers to the now-vanished South of Market and Fillmore areas, demolished to profit new development. As Walker noted, "All (…) ecologies have their counterparts in other American cities, but the peculiarities of their form, scale and survival in the Bay Area lends the place a special feeling and look" (Walker, 1995: 35). This "evidence of the coexistence" (Solnit, 2010: 115) of different experiences and unique histories, these feelings and looks, are the woven fabric of a unique form of urbanism.

In a great movement that constantly produces new forms of "being together," which negotiates and renegotiates, claims and reclaims, creates, preserves, destroys, and rebuilds, it seems that there is little place left for remembering past disasters or anticipating those yet to come. If the earthquakes have not left obvious traces, they did leave certain small, perhaps durable ones. But none of them has been consigned to a map that tells the invisible story of past disasters and risky fire corridors.

2.2.3 Entangled Inscriptions of Risk

As the Solnit recalled, "San Francisco is divided into those who remember a vanished or mutated landmark or institution, and those who came later" (Solnit, 2010: 6). In the Bay Area today, the space where earthquake risk is visible can be difficult to find. The traces of the 1906 earthquake, for instance, have almost been erased from the urban landscape. If the transformation of the landscape testifies to the impact of the momentous events of 1906, the changes that have taken place, buried under the growing vegetation and layers of stucco of successive remodeling, are hardly visible to an uninformed and inattentive eye.

Despite this lack of visibility today, the 1906 earthquake not only destroyed a good part of San Francisco, but also affected Oakland and, to a lesser extent, Berkeley. Because of the large extent of the destruction, the earthquake and fire have had long-term consequences on the urban development of the Bay Area. As refugees flew the disaster scene—some of them never to return—the economic construction boom on the Peninsula unapologetically ceased and residency patterns were redistributed to suit the scale of the Bay Area.[23] Refugees from San Francisco's "North Beach" neighborhood built homes for themselves across the Bay, in the northern parts of Berkeley. At the turn of the twentieth century, Berkeley was an area of hilly land with scrubby vegetation, but it evolved into a cozy neighborhood with large houses surrounded by luxuriant gardens, meticulously arranged. In contrast to the grid system of San Francisco, this development followed the specific topography of the Berkeley Hills (Schwartz, 2009). The picture below (Fig. 2.1) shows the spectrum of the destruction after the 1906 fire in San Francisco and Fig. 2.2 shows the extent of the reconstruction that has followed in the many decades since. The photograph in Fig. 2.1 was taken from a balloon five weeks after the 1906 earthquake; the photograph in Fig. 2.2, taken in 1996 from a similar angle, shows the contemporary urban landscape as it has been reconstructed, following a very similar grid layout.

From a distance, historic buildings have become indistinguishable from the rest of the city. Once the only visible landmarks of the past, they have become the center of the city, a core around which San Francisco has reinvented itself—stitching together past and present in such a precise way that it has become hard to distinguish one from the other. Of course, since this picture was taken, San Francisco's city line has been considerably transformed by the economic boom of the tech industry.

For the residents of the contemporary San Francisco Bay Area, the 1906 earthquake is often only known through casual conversations and stories published in the local newspapers every April 18 to commemorate the event. Occasionally, these commemorations attract more publicity, as it happened during the conference organised to pay tribute to the event

[23] "All the surrounding cities and towns are jammed with the homeless ones, where they are being cared for by the relief committees. The refugees were carried free by the railroads to any point they wished to go, and it is estimated that over one hundred thousand people have left the peninsula on which San Francisco stood" (University of Buffalo, 2008).

Fig. 2.1 The city of San Francisco after the 1906 earthquake and fire (Credit: US Geological Survey Department of the Interior/USGS U.S. Geological Survey/photo in Thatcher, Ward, Wald, Hendley, & Stauffer, 2001)

Fig. 2.2 The city of San Francisco as it looked in 1996 (Credit: US Geological Survey Department of the Interior/USGS U.S. Geological Survey/photo in Thatcher et al., 2001)

bringing together scholars and researchers with concerned residents and curious (Grossi & Muir-Wood, 2006; Grossi & Zoback, 2010; Hinman & Hutchinson, 2005; Kircher, Seligson, Bouabid, & Morrow, 2006; Perkins, Chakos, Olson, Tobin, & Turner, 2006).[24] As the scientific

[24]As we shall see later, these pieces of information are crucial for our understanding of the event. Some of this more scientific content produced for these events is accessible online, providing useful first-hand testimonies and a fuller picture of the era. Other documents, not intended to reach large audiences, have remained more confidential, even though a well-indexed Google "Advance Search" can lead one to them.

communications made explicit for the public, even a 100 years after the event, the space of the 1906 earthquake has not yet been entirely defined and new scientific elements were still being discovered as researchers and scientists developed computer models of the earthquake to grasp more precisely its unfolding and dynamic nature. Caught between denial, collective obliviousness and an always evolving science, concerned residents have nonetheless found ways to navigate the unstable, constantly shifting ground of knowledge (November, Camacho-Hübner, & Latour, 2010), and the complex web of actants that comprise it. To see the earthquake risk, one needs to follow, for instance traces of seismic activity looking more closely at the urban landscape, and following, the "signposts."[25]

2.2.3.1 Visible Trace 1. Mapping the Faults

Thanks to scientific contributions of the earthquake expert community, as well as ongoing efforts to make scientific information accessible through new interfaces of visualization, the presence and the localization of faults are now rightly part of the urban landscape. The traditional map image of the San Francisco Bay region, striated with red fault lines, has become a familiar object, memorized by many Bay Area residents.[26] Everyday tool of recognition and situation, these maps are now considered as the reference documents which summarize years of scientific research, instrumentalisation, interdisciplinary data collection and analysis.

Faults map bear the multi-dimensionality of a meeting point: as an object, they can be folded and unfolded, but as an performative artefacts, they impose a reality layering the invisible faults onto our visible environment, creating signposts that residents can learn to read. These maps are the point of interconnection of several networks, sharing the information collected by some and meant to be read and interpreted by others. The systematic collection and verification of the data through a scientific workflow provides hazard maps with an embodied "scientific conscience".[27] This instauration of the map as a performative object opens

[25] I have borrowed this word from November et al. (2010).

[26] The maps are regularly published in the newspaper and are part of the earthquake prevention material produced by the USGS and the American Red Cross, "Putting Down Roots in Earthquake Country" (American Red Cross et al., n.d.).

[27] In the case of earthquake risk in the Bay Area, a great deal of information can be found online but, as many of the informants told me, only the information produced by the USGS is considered sufficiently scientific to be reliable.

up the possibility of continuity and movement between the realms of the actors and artefacts.[28] In the same way that maps guide our journeys, the faults map allow residents of the Bay Area to "navigate" between the faults and assess the danger and create an mental representation of the space they love in.

Maps are also in communication with the visible built environment. Following the traces of the faults, and although the presence of earthquake risk can seem remote, approaching certain buildings the potential threat suddenly becomes more potent. The presence of the active fault is becomes a natural intangible heritage, one actant among others that forms the specific Californian ecosystem, its geodiversity.

For inhabitants of the East Bay, the space of risk is, in fact, a space of multiple dangers. The Hayward Fault runs just below the hills in Berkeley and Oakland, which are subject to sudden and hard-to-control wildfires, as well as potentially dreadful landslides. Something that residents are well aware of: "Where I live, the biggest risk is earthquakes because I live close to the Hayward Fault. We live on a hillside, so the other risk is the landslides" [S.16]. Here, the traces of seismic activity are visible to those who know where to look—at one of the most iconic and controversial buildings on the UC Berkeley campus: the California Memorial Stadium, or at random places where the curve as been distorted by continuous pressure of the tectonic movement.

[28] "But what we just said about the spurious distinction between 'physical' and 'human' geography is even truer of the efforts to add the fourth dimension to the 'three dimensions' of Euclidian space. To be sure, once you believe you have frozen the navigational movements in the three dimensions of Euclidian space, it is very difficult to see how you could insert the obvious fact of movement and transformation. But this difficulty vanishes once you realize that in geography – provided you shift to the navigational interpretation of maps – everything is on the move: the navigator in the yacht, the yacht itself, the pencil on the map, the tide, the current, the Nautical Service in charge of sinking the buoys, in brief the whole damned multi-verse. The very idea of a time separated from a space (as if a fourth dimension had to be added to the three of 'commonsense' as if living in a Euclidian space was commonsense!) comes from dreaming over a map too long. [...] The very idea of a mobile moving without undergoing any transformation is the result of an aesthetic contemplation of an isolated inscription (Latour, 1986). "It is not a property of the world at least not of the multiverse" (November et al. 2010: 596).

2.2.3.2 Visible Trace 2. The California Memorial Stadium

In 1998, the *San Francisco Chronicle* created a sensation when it pointed out that the stadium, which could seat as many as 71,799 people, and occasionally did during college football season, had "become the biggest risk of massive loss of life from an earthquake in the U.S., if not the world" (Burress, 1998).

Since its construction, the Berkeley stadium has been a controversial subject among students, staff, and residents: indispensable to the prestige of the University, it was built in 1923 and has been considered a landmark ever since. At the time of its construction, the project was strongly contested by residents of the Berkeley Hills[29] as it was erected in a canyon considered one of the ecological jewels of the East Bay and known for its geological feature, resulting of seismic activity along the Hayward Fault.

Despite the initial opposition, the stadium remained a fixture of student life for decades. But as soon as the renovation project started, opposition resurfaced. Between December 2, 2006, and September 9, 2008, "tree sitters" opposed the destruction of 90 trees. During this time, the tension became especially tense when the construction site was revealed to be a burial site and sacred place for Olhone Native Americans (Jones, 2007). Finally, the University project was a bombshell for the greater Berkeley bohemian community, who had been landscaping the area with a great sense of natural preservation.[30] Despite these strong challenges, the stadium was built and remains today (Fig. 2.3).

In 2008, engineers in charge of retrofitting the stadium said they had solved "one of the world's great retrofit puzzles," and the lead engineer claimed, "I'll sleep well at night, even if I have season tickets in Section KK" (Jones, 2008). The solution consisted of slicing portions of the

[29] Among the opponents to the initial construction project and its retrofitting were several environmental supporters, including several board members of the Sierra Club. According to Daniella Thompson, some members of the Sierra Club had built their homes in Berkeley Hills in the late nineteenth and early twentieth centuries. Other residents there included scholars of the University of California. John Muir, the President of the Sierra Club, who visited the place, described it as "a most glorious season of terrestrial grace" (http://berkeleyheritage.com/eastbay_then-now/sierra_club_leaders.html).

[30] "Community power and design with nature suffered another blow when the aesthetics of football triumphed at the University and a huge stadium was jammed into scenic behind the campus Strawberry Canyon" (Walker, 1995).

Fig. 2.3 A gap caused by movement in the Hayward Fault after the retrofitting work

stadium into "blocks that will rest on plastic sheets."[31] But not everybody was convinced: scientists and activists came back into the debate. In a letter to the Regents of the University of California, Berkeley, Professor Emeritus Garniss H. Curtis (1919–2012) of the Department of Earth and Planetary Science stated:

> As the request for my geologic opinion on the advisability of constructing large buildings in the lower part of Strawberry Canyon and in the next canyon to the north known as Blackberry Canyon came to me on May

[31] "The plan, which is expected to get under way in the next year or two, calls for portions of the stadium to be sliced into blocks that will rest on plastic sheets. When the earth ruptures, the soil will move under the sheets but, engineers hope, will leave the blocks intact. The price tag for the retrofit is estimated at between $150 million and $175 million" (Jones, 2008).

4th, I have to be brief and rely on my memory. I shall first say as strongly as I can "absolutely do not construct any buildings in those two canyons". (Curtis, 2008)[32]

Despite the opposition of activists and advocates, of scientists and political entities, of descendants of buried Native Americans and living students, UC Berkeley finally evicted the "tree sitters" and started their retrofitting work. The UC Board of Regents had already approved $321 million for the project. The stadium was reopened in September 2012, just in time for the school's football season.

2.2.3.3 Visible Trace 3. The Bridges

Acknowledgement of the "concreteness," the specific materiality of risk, is vivid for commuters, who actively use the Bay Area's civil engineering structures each day. Most of the bridges and tunnels are known by both residents and experts to be particularly vulnerable to earthquakes.[33] "I drive across the bridge a lot, and I live in the city [San Francisco], so my biggest risks are fire and earthquake", said one respondent [M.3]. This fear of bridges is rooted in the history of the Bay. The Cypress Street Viaduct, part of the Nimitz Freeway,[34] which used to cross through some of the low-income neighborhoods in Oakland, partly collapsed during the 1989 Loma Prieta earthquake. That day, the upper tier fell on top of the lower one, resulting in 42 casualties (Fig. 2.4).

The Nimitz Freeway was opened to the public in 1957 and, ever since, it had been criticized for spatially segregating an already underprivileged area of Oakland from the more affluent parts of the city. Additionally, there had been awareness of the vulnerability of the actual structure since seismic retrofitting work had been undertaken following

[32]The reason for his strong opposition was the high risk of landslides and seismic activity.

[33]From north to south, the bridges are: the Carquinez Bridge, the Richmond and San Rafael Bridge, the San Francisco-Oakland Bay Bridge, the San Mateo Bridge, the Dumbarton Bridge, and, of course, the Golden Gate Bridge. As we shall see later, despite the effort of Bay Area Rapid Transit authorities, experts estimate that the underwater tunnel crossing underneath the San Francisco Bay cannot be totally retrofitted; consequently, it may not be entirely safe during a major earthquake.

[34]Some people also discussed the Bay Area Rapid Transit public transportation system tunnel, which runs underneath the Bay, joining localities in the East Bay to the San Francisco Peninsula.

Fig. 2.4 Side view of support-column failure and collapsed upper deck, Cypress Viaduct (Credit: US Geological Survey Department of the Interior/USGS U.S. Geological Survey/photo by H.G. Wilshire)

the 1971 San Fernando earthquake. During one interview, a respondent recalled:

> It was known that the freeway might collapse from the '60s and '70s because of two reasons. One, the design, based on future information, it was inadequate. There may also have been some malfeasance in the concrete itself; in other words, the contractors may have cheated on the concrete. They also may not have put in reinforcement bars correctly. They did know the design of the freeway might lead to a high likelihood that it may collapse. They knew that for 20 years … if you don't put enough steel in, then it doesn't help the concrete. If it's not the right mixture and uses lesser quality of cement, it might collapse anyways. They finally declared it to be okay, but what does that mean? I don't know. The contractor cheated, and what's done is done, are you going to go and jackhammer in there? The state would have to come up with more money to do that. […] As far as I know they never tested the materials. They didn't. Probably wasn't the only freeway in California where they didn't. It falls down, 49 [*sic*] people died and California pays all the victims, we don't know how

much because there's a clause in the payment saying they can't divulge that information, the state quieted it up.[35] [A.28]

After the 1989 earthquake, thanks to the mobilization of West Oakland residents the structure was torn down and a landscaped boulevard was designed to recreate the links between the different parts of Oakland.

Today, the Mandela Parkway stands in place of the former freeway, and locally owned restaurants and Sunday vegetable markets have given a new face to the space. Even so, certain stubborn memories remain anchored in the minds of those who witnessed the catastrophe, even long after material traces of the collapsed bridge were removed from the landscape:

> That Cypress Freeway, as that overpass was called, was a monumental structure in the most literal sense, a gray mass of concrete as high as an office building, and when I first got there in the early evening I remember also that I looked up and felt dizzy and sick, as though I were staring at a body that had been disemboweled. The ripped freeway had opened a great cross-section of gigantic construction innards that sprawled and jutted and smoked and hissed. Parts of cars were visible inside the striations of concrete and metal, but the whole arrangement made no sense; you couldn't understand what was roadway and what was broken piling, and on the ground were cars upside down and jerked over at weird angles between the fire trucks and improvised emergency equipment, and at one spot below the overpass a crowd of medical workers had gathered at the base of a long ladder that stretched right up into the mess overhead. (Gorney, 2009)

The full impacts of such a tragedy, both in the long and short terms, are difficult to evaluate. They affect both the landscape and the people living in it. Despite the enormous mass of debris left in the middle of Oakland, the consequences of the destruction of the highway's upper and lower decks travelled through space. The space of transit became a space of death. Everything was transformed.

[35] The most visible and long-term consequence of the Loma Prieta earthquake was the construction of a new segment of the San Francisco Bay Bridge east span, which began in 2002. As did the retrofitting operations of the Memorial Stadium, this construction has triggered heated controversies and debates.

> The coroner's list was like a census report on the polyglot Bay Area: victims were Asian, black, white, Hispanic. There was a Palestinian, a refugee from Vietnam, Midwesterners, Easterners. Most were not native to the Bay Area: they had come for economic opportunity, for personal or political freedom, for the climate, the beauty. (Reinhold, Navarro, & Rabinovitz, 1989)

Some of the victims were regular commuters; others just happened to be driving by. In a matter of moments, their existences—and, with them, their movements and connections to other worlds, other people, goals, and drives—stopped. The role of chance, or fate, or bad luck, with this sudden interruption by the earthquake, created a massive chain of consequences for innumerable people, adding an important dimension to the narrative of the catastrophe:

> She [Donna Marsden] boarded the blue and gray Dodge van every weekday morning, and again at 4:40 P.M. in San Francisco for the return home. That day had been no different, except that her husband had taken her to lunch to celebrate her promotion at the University of California at San Francisco. Later, the van traveled the upper deck of the Nimitz Freeway at almost the same time it did every day. Had the quake struck seconds earlier or later, Mrs. Marsden would scarcely have been affected, apart from picking up some fallen items at home. (Reinhold et al., 1989)

During the rescue operations inside the debris, time and space, life and death, risk and security no longer had the same contours that they used to. Usual objects had to be used in unusual ways; hard decisions had to be made.[36]

The event was broadcast widely, showing the extremely difficult working conditions of the rescue teams the graphic images of the consequences of the catastrophe on the structure of the bridge—which included not only crushed cars, but also human bodies. These images were shared by many: neighbors, families, friends, and, of course, a large,

[36] One of the most dramatic stories was the rescue of a young boy, who endured the amputation of his legs in order to free his body from concrete blocks: "For the journalist who had covered the story, details of the rescue of this young boy are still vivid: 'I remember Betts emerging from the operating room twenty years ago to talk to us; it was 4:30 in the morning by that time, and he was in his scrubs, unshaven, gray around the eyes, and worried about shock and toxins that might be released into the boy's lungs or kidneys'" (Gorney, 2009).

distant television audience.[37] A large community of witnesses had to deal with direct or indirect memories of this terrible event. Indeed, the emotional consequences for the rescue teams have been largely documented, as some have endured distressing and recurrent memories of the experience in the years since:

> During the rescue work on the collapsed highway, firefighters worked in the pancaking space between the upper and lower levels of the freeways. [...] There were strong aftershocks and subsequent compression of the levels. As the space compressed, the escape routes became smaller, increasing the risk that workers would become trapped. Gasoline was leaking from crushed vehicles, increasing the risk of explosions if an extrication tool caused sparks. Personnel were exposed to traumatic stimuli, hearing cries for help became dim as time passed, or witnessing extraordinary procedures such as necessary limb amputation with chainsaw to extricate one of the person trapped. (Myers & Wee, 2005: 185)

Looking for the traces of risk in the urban landscape opens up the deployment of actants, which contribute to making the complex phenomenon of risk construction visible. These stories of destruction draw a moving definition of the risk. In looking for these traces, we find multiple actants—steel, concrete, contractors, scientific proof, the State, money, information, responsibility, injuries, death, trauma, heroism, malfeasance, miserliness—all engaged in a complex activity that involves disappearance, recalcitrance, appearance, and the transformations of space, objects, memory, and people.

Not all the dimensions of this network can be grasped at first, and not all of them move in the same direction. Some of them are measured, planned, and calculated—a commemoration, or the construction of a new bridge, for example. Some meet the stated aim of a particular need: they are part of the controversies; they are involved in moral questions, and monetary valuations. Others are messy, uncontrolled, and can fall into the spectrum of attention, such as memories and feelings. These latter aspects are probably the least visible in the overall landscape; they are

[37] "I was living in Maine. I once lived in San Francisco and was galvanized by the images on TV of familiar places in ruins. So yes, shock waves were far reaching. San Francisco is one of those "national cities" that everyone identifies [with] – had it been someplace like Topeka or Omaha, the images of ruin would not have carried the shock of recognition in the rest of the country," recalls a respondent [J., 2009].

the least easy to capture with scientific tools, but still they exist and, as such, they shape many different spaces and subjects.

In the recent history of California, the status of risk as an "object," or a "thing" in a Souriaun sense, is an ongoing negotiation. These risks are a negotiation between actants, some of which are visible and some not; some standing along, some combined; some strong, others weak, all present in their phantasmagoric existence. As Solnit described it: "I also live among ghosts. For the better or worse, the familiar vanishes, so that the longer you live here, the more you live with a map that no longer matches the actual terrain. After the great 1972 earthquake, Managua, Nicaragua, lost many of its landmarks; people long after gave directions by saying things like, 'Turn left where the tree used to be'" (Solnit, 2010: 6). Things are hidden, buried, contested, and denied; they are reduced to the potentiality of an expression not fully expressed, but somehow, they—at least partially—remain. As a consequence, the risk, as a reality shared by many, has a hard time to exist fully. To look for the traces of seismic activity requires the excavation of past histories, the digging trenches, the studying of walls, and seeking out the prints left by past disasters.

References

Amador, F. (2004). The causes of 1755 Lisbon earthquake on Kant. *Actas VIII Congreso de La Sociedad Espanola de Historia de Las Ciencias Y de Los Tecnicos*, 485–495. Retrieved from http://dialnet.unirioja.es/servlet/dcfichero_articulo?codigo=1090088.

American Red Cross, B. A. C., Association of Bay Area Governments, California Earthquake Authority, California Geological Survey, Earthquake Engineering Research Institute, Governor's Office of Emergency Services, ... U.S. Geological Survey. (n.d.). *Putting down roots in earthquake country—Your handbook for the San Francisco Bay region*. Retrieved July 25, 2012, from http://pubs.usgs.gov/gip/2005/15/.

Beck, U. (1992). *Risk society, toward a new modernity* (M. Ritter, Ed.), *Nation* (Vol. 2). Sage. https://doi.org/10.2307/2579937.

Beck, U. (2008). *World at risk. New York* (Vol. The Report). Polity Publisher. Retrieved from http://scholar.google.com/scholar?hl=en&btnG=Search&q=intitle:World+at+risk:+The+report+of+the+Commission+on+the+Prevention+of+WMD+Proliferation+and+Terrorism#0.

Beck, U. (2009). Critical theory of world risk society: A cosmopolitan vision. *Constellations, 16*(1), 3–22.

Berglund, B. (2007). *Making San Francisco America: Cultural frontiers in the urban west, 1846–1906*. Lawrence: University Press of Kansas.

Bessy, C., & Chateauraynaud, F. (1995). *Experts et Faussaires*. Paris: Métailié.

Bloom, J. (2014). *An early warning device of earthquakes (and other maladies) for everyone*. Retrieved from http://5nf5.blogspot.com/2014/09/early-warning-device-of-earthquakes-and-other-maladies-for-everyone.html.

Borraz, O. (2008). *Les Politiques du Risque* (Sciences P). Paris: Les Presses de Sciences Po.

Bourg, D., Joly, P.-B., & Kaufmann, A. (2013). *Du risque à la Menace* (D. Bourg, P.-B. Joly, & A. Kaufmann, Eds.) (Ecologie e). Paris: Presses Universitaires de France.

Brechin, G. (2006). *Imperial San Francisco, urban power, earthly ruin*. Berkeley: University of California Press.

Burress, C. (1998). *Quaking at Berkeley—New study of Northern Hayward fault forces UC to revise damage estimates read more*. Retrieved from http://www.sfgate.com/news/article/QUAKING-AT-BERKELEY-New-study-of-northern-3008635.php#ixzz2NyfDL5Cz.

Burton, I., Kates, R., & Whites, G. (1993). *The environment as hazard* (2nd ed.). New York: The Guilford Press.

Cabane, L. (2012). *Gouverner les catastrophes. Politiques, savoirs et organisation de la gestion des catastrophes en Afrique du Sud*. CSO.

Callon, M., Lascoumes, P., & Barthe, Y. (2009). *Acting in an uncertain world: An essay on technological democracy*. Cambridge: MIT Press.

Charvolin, F. (1997). La Naissance du Domaine de l'Environnement. *Strates [Online], 9*. Retrieved from http://strates.revues.org/636.

Clark, L. (2006). *Worst cases, terror and catastrophe in the popular imagination*. Chicago: University of Chicago Press.

Clavandier, G. (2004). *La mort collective* (CNRS Socio). Paris: CNRS Editions.

Coen, D. (2013). *The Earthquake observers: Disaster science from Lisbon to Richter*. Chicago: The University of Chicago Press.

Collectif. (2008a). Catastrophe(s)? *Le Portique, 2ème semes*(22).

Collectif. (2008b). *Le temps des Catastrophes* (M.-O. Padis, Ed.), *Esprit* (Esprits, Vol. Mars–Avril).

Creton-Cazanave, L. (2010). *Penser l'Alerte par les Distances. Entre Planification et Emancipation, l'Exemple du Processus d'Alerte aux Crues Rapides sur le Bassin Versant du Vidourle*. Université de Grenoble.

Curtis, G. H. (2008). *Emeritus professor of geology weighs in on memorial stadium*. The Berkeley Daily Planet.

Davis, M. (1998). *Ecology of fear: Los Angeles and the imagination of disasters*. New York: Vintage Books.

Deleuze, G. (1987). *Leibniz. Logique de l'évènement*. Retrieved from www.web-deleuze.com.

Demuth, J. L. (2002). Countering terrorism: Lessons learned from natural and technological disasters. In *National Academy of Sciences* (p. 36). Washington, DC.

D'Ercole, R., Gluski, P., Hardy, S., & Sierra, A. (2009). Vulnérabilités urbaines dans les pays du sud. *Cybergeo: Journal of Geography, 1,* 95–98.

Dupuy, J.-P. (2002). *Pour un catastrophisme éclairé. Quand l'impossible est certain* (Point Essa). Paris: Seuil.

Dynes, R. R. (1997). *The Lisbon earthquake in 1755: Contested meaning in the first modern disaster* (No. #255).

Favier, R., & Granet-Abisset, A.-M. (2009). Society and natural risk in France, 1500–2000: Changing historical perspectives. In C. Mauch & C. Pfister (Eds.), *Natural disasters cultural response* (pp. 103–136). Lanham: Lexington Book.

Fressoz, J. (2012). *L'apocalypse joyeuse: Une histoire du risque technologique.* Paris: Seuil.

Gaar, G., & Miller, R. W. (2006). *San Francisco: A natural history* (Image of A). San Francisco: Arcadia Publishing.

Gorney, C. (2009). *Remembering the surgery inside the freeway collapse.* Retrieved from http://oaklandnorth.net/2009/10/17/remembering-the-surgery-inside-the-freeway-collapse/.

Gralepois, M. (2008). *Les risques collectifs dans les agglomérations françaises. Contours et limites d'une approache territoriale de prévention et de gestion des risques à travers le parcours des agents administratifs locaux.* Universite de Paris Est.

Grossi, P., & Muir-Wood, R. (2006). *The 1906 San Francisco earthquake and fire: Perspectives on a modern super cat.* San Francisco.

Grossi, P., & Zoback, M. Lou. (2010). *1868 Hayward earthquake: 140 year retrospective.* Menlo Park: Risk Management Solutions.

Groth, P., & Bressi, T. (1997). *Understanding ordinary landscapes.* New Haven: Yale University Press.

Haas, P. (1992). Epistemic communities and international policy coordination. *International Organization, 46*(1), 1–35.

Hackett, E., Amsterdamska, O., Lynch, M., & Wajcman, J. (Eds.). (2008). *The handbook of science and technology studies.* Cambridge: MIT Press.

Hansson, S. O. (2012). Risk. In E. N. Zalta (Ed.), *The Stanford encyclopedia of philosophy.* Stanford: Stanford University. http://www.plato.stanford.edu/entries/risk/.

Haraway, D. (1988). Situated knowledges: The science question in feminism and the privilege of partial perspective. *Feminist Studies, 14*(3), 575–599.

Henry, D. (2005). Anthropology and disasters. In D. McEntire & W. Blanchard (Eds.), *Disciplines, disasters and emergency management: The convergence and divergence of concepts, issues and trends from the research literature* (p. 1). Federal Emergency Management Agency. https://doi.org/10.2307/3032499.

Hinman, E., & Hutchinson, D. A. (2005). *San Francisco's earthquake risk.*

Hoffman, S. (1994). Up from the embers: A disaster survivor's story. *NCP Clinical Quarterly, 4*(2), 15–16.

Hoffman, S. (1998). Eve and Adam among the embers: Gender patterns after the Oakland Berkeley firestorm. In E. Enarson & B. Hern Morrow (Eds.), *The gendered terrain of disaster: Through women's eyes* (pp. 55–61). Westport, CT: Greenwood Publishing.

Hoffman, S. M., & Oliver-Smith, A. (2002). *Catastrophe and culture: The anthropology of disasters* (S. M. Hoffman & A. Oliver-Smith, Eds.). School of American Research Press and James Currey.

Houdart, S. (2008). *La cour des miracles: Ethnologie d'un laboratoire japonais* (Sociologie). Paris: CNRS Editions.

Houdart, S., & Thiery, O. (2011). *Humains non humains. Comment repeupler les sciences sociales?* Paris: La Découverte.

Ingold, T. (2005). *Learning through doing and understanding in practices.* Retrieved from http://www-new1.heacademy.ac.uk/assets/Documents/subjects/csap/Learning_though_doing_and_understanding_in_practice.pdf.

Issues Answer. (2009). *San Francisco Bay Area preparedness—executive summary.*

James, W. (1906). On some mental effects of the earthquake. The Youth's Companion. Reprinted in James, H. Jr., 1911, *Memories and studies* (H. James, Jr.). Longmans, Green, & Co. (Vol. June). New York.

Jasanoff, S. (1986). *Risk management and political culture* (Social Res). New York: Russell Sage Foundation.

Jasanoff, S. (2010). Beyond calculation: A democratic response to risk. In *Disaster and the politic of intervention* (pp. 14–41). New York: Columbia University Press.

Jonas, H. (1995). *Le principe de responsabilité* (Champs Ess). Paris: Flammarion.

Jones, C. (2007, February 20). Tree-sitters say site near stadium may be Native American burial ground. *San Francisco Chronicle.* Retrieved from http://www.sfgate.com/bayarea/article/BERKELEY-Tree-sitters-say-site-near-stadium-may-2647672.php.

Jones, C. (2008, September 25). Retrofit plan to ride out quake at Cal stadium. *SFGates.* Retrieved from https://www.sfgate.com/bayarea/article/Retrofit-plan-to-ride-out-quake-at-Cal-stadium-3193674.php.

King, J. (2011). *Cityscapes: San Francisco and its buildings.* Berkeley: Heyday.

Kircher, C. A., Seligson, H. A., Bouabid, J., & Morrow, G. C. (2006). When the big one strikes again—Estinated losses due to a repeat of the 1906 San Francisco earthquake. *Earthquake Spectra, 22*(2), 297–339.

Knobel, L. (2014). Cal prof Joshua Bloom makes earthquake alert kit for $110. Retrieved from http://www.berkeleyside.com/2014/09/09/uc-berkeley-prof-joshua-bloom-makes-earthquake-alert-kit-for-110/.

Lagadec, P. (1987). Le risque technologique majeur. In *Universalia*. Paris: Encyclopedia Universalis.

Langumier, J. (2008). *Survivre à l'innondation. Pour une ethnologie de la catastrophe*. Lyon: ENS éditions.

Lash, S., Szerszynski, B., & Wynne, B. (1996). *Risk Environment and Modernity: Toward a New Ecology* (L, Ed.) (Published). London: Sage.

Latour, B. (1991). *Nous n'avons jamais été modernes. Essai d'anthropologie symétrique* (Poche). Paris: La Découverte.

Latour, B. (2004). Whose cosmos? Which cosmopolitics? A commentary on Ulrich Beck's peace proposal? *Common Knowledge, 10*(3), 450–462.

Latour, B. (2013). *An inquiry into modes of existence: An anthropology of the moderns*. Cambridge, MA: Harvard University Press.

Lupton, D. (1999a). *Risk*. London: Routledge.

Lupton, D. (1999b). *Risk and sociocultural theory: New directions and perspectives*. Cambridge: Cambridge University Press.

Lussault, M. (2007). *L'homme spatial, la construction sociale de l'espace humain*. Paris: Editions du Seuil.

Mialet, H. (2012). *Hawking incorporated: Stephen Hawking and the anthropology of the knowing subject*. Chicago and London: University of Chicago Press.

Miller, F., Osbahr, H., Boyd, E., Thomalla, F., Bharwani, S., Ziervogel, G., … Nelson, D. (2010). Resilience and vulnerability: Complementary or conflicting concepts? Retrieved from http://www.ecologyandsociety.org/vol15/iss3/art11/.

Mitchell, T. (2002). *Rule of experts: Egypt, techno-politics, modernity*. Berkeley: University of California Press.

Montgomery, D. (2004). *King of fish: The thousand year run of salmon*. New York: Basic Books.

Myers, D., & Wee, D. F. (2005). *Disaster, mental health services: A primer for practitioners*. New York: Brunner-Routledge.

Neyrat, F. (2008). *Bio-politique des catastrophes* (Dehors). Paris: Editions MF.

November, V., Camacho-Hübner, E., & Latour, B. (2010). Entering a risky territory: Space in the age of digital navigation. *Environment and Planning D: Society and Space, 28*(4), 581–599. https://doi.org/10.1068/d10409.

November, V., Penelas, M., & Viot, P. (2009). When flood risk transforms a territory: The Lully effect. *Geography, 94*, 189–197.

Oliver-Smith, A., & Hoffman, S. M. (1999). In A. Oliver-Smith & S. M. Hoffman (Eds.), *The angry earth: Disaster in anthropological perspective*. Oxford, UK: Routledge.

Pattaroni, L., & Pedrazzini, Y. (2010). Insecurity and segregation: Rejecting an urbanism of fear. In P. Jacquet, R. Pachauri, & L. Tubina (Eds.), *Cities: Steering towards sustainability* (pp. 177–187). Delhi: TERI Press.

Pelling, M. (2003). Tracing the roots of urban risk and vulnerability. In *The vulnerability of cities: Natural disasters and social resilience* (p. 224). London: Routledge.
Perkins, J. B., Chakos, A., Olson, R. a., Tobin, L. T., & Turner, F. (2006). A retrospective on the 1906 earthquake's impact on Bay Area and California public policy. *Earthquake Spectra*, *22*(S2), S237. https://doi.org/10.1193/1.2181527.
Quarantelli, E. L. (1999). *Disaster related social behavior: Summary of 50 years of research findings.*
Quarantelli, E. L., & Perry, R. (2005). *What is a disaster*. Xlibris: Bloomington.
Quenet, G. (2005). *Les Tremblements de terre aux dix-septième et dix-huitième siècles. La naissance d'un risque*. Seyssel: Champ Vallon.
Reghezza-Zitt, M. (2013). *Utiliser la polysémie de la résilience pour comprendre les différentes approches du risque et leur possible articulation*. EchoGéo [En Ligne], 24. https://doi.org/10.4000/echogeo.13401.
Reinhold, R., Navarro, M., & Rabinovitz, J. (1989). *The freeway dead: Portraits from Oakland—A special report. 11 whose lives ended as quake crushed I-880*. Retrieved from http://www.nytimes.com/1989/10/29/us/freeway-dead-portraits-oakland-special-report-11-whose-lives-ended-quake-crushed.html?pagewanted=all&src=pm.
Reisner, M. (2003). *A dangerous place*. New York: Pantheon Books.
Revet, S. (2007). *Anthropologie d'une catastrophe. Les coulées de boue de 1999 au Venezuela*. Paris: Presses Sorbonne Nouvelle.
Schwartz, R. (2009, December). Eccentrics, heroes and cutthroats of old Berkeley. *SF Gate*. Retrieved from http://www.sfgate.com/bayarea/article/Eccentrics-Heroes-and-Cutthroats-of-Old-Berkeley-2523725.php#photo-2060993.
Serres, M. (2009). *Temps des crises* (Manifestes). Paris: Le Pommier.
Solnit, R. (2010). *Infinite city: A San Francisco atlas*. Berkeley: University of California Press.
Star, S. L., & Griesemer, J. (1989). Institutional ecology, "translation", and boundary objects: Amateurs and professionals in Berkeley's museum of vertebrate zoology, 1907–1939. *Social Studies of Sciences*, *19*(3), 387–420.
Stengers, I. (2009). *Au temps des catastrophes; résister a la barbarie qui vient* (les empêch). Paris: La Découverte.
Thatcher, W., Ward, P., Wald, D., Hendley, J. I., & Stauffer, P. (2001). *When will the next great quake strike Northern California?*
Tierney, K. J. (1994). *Sociology unique contributions to the study of risk* (No. #204).
Tobriner, S. (2006). *Bracing for disaster: Earthquake-resistant architecture and engineering in San Francisco, 1838–1933*. Berkeley: The Bancroft Library, University of California Berkeley.

Tulloch, J., & Lupton, D. (2003). *Risk and everyday life*. London: Sage.
University of Buffalo. (2008). *Jack London and the 1906 earthquake*. Retrieved from http://mceer.buffalo.edu/1906_Earthquake/personal_stories/jack-london.asp.
VonWinterfeldt, D., Roselund, N., & Kitsuse, A. (2000). *Framing earthquake retrofitting decisions: The case of hillside homes in Los Angeles*.
Walker, R. (1995). Landscape and city life: Four ecologies of residence in San Francisco Bay Area. *Cultural Geographies, 2*(1), 33–64. Retrieved from http://oldweb.geog.berkeley.edu/ProjectsResources/Publications/BayLandscape.htm.
Walker, R. (1998). An Appetite for the City. In J. Brook, C. Carlsson, & N. J. Peters (Eds.), *Reclaiming San Francisco, history, politics, culture* (p. 355). San Francisco: A City Lights Anthology.
Walker, R., & Thomas, S. (2010). Blinded by history: The geographic dimension of environment and society. In D. Cazaux Sackman (Ed.), *A companion to American environmental history* (pp. 553–578). Blackwell Publishing. https://doi.org/10.1002/9781444323610.
Walter, F. (2008). *Catastrophes. Une histoire culturelle XVI-XXIème siècle*. Paris: Editions du Seuil.
Watts, M. J., & Bohle, H. G. (1993). The space of vulnerability: The causal structure of hunger and famine. *Progress in Human Geography, 17*(43), 43–67. https://doi.org/10.1177/030913259301700103.
Wiener, J. B. (2010). *The real pattern of precaution*. Washington, DC: Future Press.
Wiener, J. B., & Rogers, M. D. (2002). Comparing precaution in the United States and Europe. *Journal of Risk Research, 5*(4), 317–349. https://doi.org/10.1080/1366987021015368.

CHAPTER 3

Traumatic Legacies: Shaping the Space of Risk

3.1 Controversies

Controversies are part of the framing of risks and disasters, and consensus is often hard to obtain. This chapter examines two disasters that have affected to the Bay Area in the course of the twentieth century: the 1906 earthquake and fire and the 1991 Oakland hills fires. As we've mentioned, in the case of the 1906 event, the discussions had focused defining the respective natures and scales of the damages caused by the earthquake and the fire. The questions that followed the Oakland Fire that killed 25 people and burned 3354 houses in 1991 are of different nature. They concern the relevance of the event itself, and its definition as "a disaster." In relation to the geographical scale of the Bay Area, the perimeter of the fire was relatively small. In the following excerpt, the respondent uses a provocative, figure-based approach to explain his view about the size of the event.

> The Oakland Hills fire affected 3,000 housing units. There are 13 million people in the greater Bay Area. For most of them it is meaningless! It was just a house fire and it really has no effect on [other residents] and does not enter the active memory. [...] You know, we just don't have an event

C. Mazel-Cabasse, *Waiting for the Big One*,
https://doi.org/10.1007/978-3-030-15289-5_3

> of this magnitude in the U.S. in recent memory. Katrina is the only event in my mind that qualifies as memorable. [M.3][1]

This statement was, of course, contested by fire survivors, who claimed they "have disasters a lot" [S.16], that the fire actually affected the Bay Area as a whole; and people have held onto memories of the event.

> People, over in Marin, people down the Peninsula were seeing ashes in the air! It affected everybody, really a much broader circle than just the immediate East Bay. I still run into people now who talk about where they were the day of the fire. I say: "I lost my house" and they say: "Oh, my friend up [in] the hills too!" Everybody has a story about it. It really did affect everybody. [S.16]

Question of definition brings together not only questions of border, scale, and distance, but also the personal experiences of those borders, scales, and distances. Disasters do not happen in a vaccum: they happen at specific times and locations. Like many American cities, Oakland suffers from spatial, economic, and racial segregation.[2] These inequalities are still very vivid today, both on the maps and in the minds of the local residents.

[1] After the quotation, the discussion continued as follows:

"C: What about 9/11?

M: Whoa…5000 people, it was not a lot of people! It has a psychological impact in the fact that Americans felt that their country was impacted, but that is very different from the very impact of the disaster. The real impact of the disaster was a tiny, little part of Manhattan. I mean, nobody lost power across the river, you know. It was a very small number of people. There was inconvenience relative to the size of Manhattan and the size of [New York]. And so those things, they don't affect people on the same level. We did not have an earthquake of the scale of the one [that] affected China, where you have 5 million people displaced, or Haiti where you have 2 million people displaced. We hadn't had this sort of disaster in the U.S." (Interview, M., 2009).

[2] Self (2005) has shown that the construction of the city itself is entrenched between conflicting movements that emerged from the progressive suburbanization of the city: defensive Black Power community politics on one side, and, on the other, the White conservative homeowner associations. As Self noted, "In 1945 boosters in Oakland envisioned their metropolis in the tradition of California Urbanist, as a verdant, interconnected garden that combined suburban growth and urban vitality. Thirty years later the Black Panthers Party and other African American activists viewed Oakland as an exploited colony that was controlled from the suburban perimeter" (Self, 2005: 4).

This specific context plays a role in the second controversy which concerns the socioeconomic categories of populations affected by the fire.[3] The fire happened in upper-middle class neighborhoods and, as such, residents were able to mobilize a considerable amount of resources to rebuild, in a relatively short amount of time large and comfortable houses.[4] A privilege that has been acknowledged by the fire survivors: "My older daughter said to me: 'Gee, Mom, that makes us like the homeless, doesn't it?' I go…'Yes'… and then she said, 'But we still have money, so we'll be okay'. I said 'Yes'" [S.16]. The economic dimension, like having fire insurance, was certainly a major component in the survivors' recovery process:

> We went to some meetings; there were some town hall meetings about how to deal with your insurance company. We didn't have any of those issues that some people have. My husband is an attorney, he's used to talking to insurance companies. We had a sit-down meeting with them at his office in his conference room on his turf with our architect and our contractor, we went through [it all] item-by-item, and we had very smooth time dealing with them. We were among the first to start rebuilding. [R.5]

But, of course, not all the reconstruction stories were that straightforward and for many residents the experience of reconstruction was another layer of pain and difficulty, one that for many lasted for years.

Sometimes actants can be hard to trace: they hide behind closed doors, between the lines of the stories, and deep in the recollections of memories. Most of the time, they remain visibly unseen; rather, they need to be felt and heard. Their manifestations mobilized a range of transitory, seemingly incomplete, and hybrid artefacts, absent from scientific reports. Despite being concealed, and partly discredited, the 1991 fire transformed the space of the Oakland Hills into a succession of different assemblages: the fire itself, the ruins, and, ultimately, the rebuilt neighborhood. Work of recollection and documentation brings a fascinating body of insights, allowing us to follow the eruption of the fire, almost hour by hour.

On a warm Sunday morning of October 1991, around 10:00 a.m., the wooden houses, the Indian summer, the imported vegetation and

[3] Building up on Mike Davis' essay, "The Case for Letting Malibu Burn" (Davis, 1998).

[4] Accomplished via financial remuneration by the homeowners' insurances companies.

beloved gardens, the neighbors, the charming winding streets, the cars, and the fresh pools, were swiftly replaced by a blaze, residential fires, traffic jams, unbearable temperatures, and exhausted firefighters. The documentation available provides evidence of the healing powers of narration and art: photography, documentaries, novels, children's books, reports, scientific and journalistic articles, and written notes all create a patchwork of experiences and perspectives that retrace the transformative power of the fire. In this material, I include excerpts from face-to-face interviews conducted during my field research.[5]

3.1.1 The Oakland Fire Traces

For those who are discovering the Oakland Hills for the first time, the neighborhood looks like any recently built, affluent area. Contemporary houses, replica styles of traditional ranch or Tudor, Japanese, or Mediterranean architecture seem to have been indistinctly juxtaposed one after the other, right to the very edges of their lots. The narrow, winding roads going up and down the steep hillsides break the monotony of the seemingly overbuilt area. At street level, little is offered to the view except the multi-car garage doors and, sometimes, professionally designed landscaping. In contrast, when turning toward the west, the direction in which all the façades face, there appear the magnificent views of the Bay, the Oakland harbor, the bridges, and the city of San Francisco. Except for the "Fire Storm Memorial Garden," which sits alongside the highway, again, traces of what happened in 1991 are hard to find.

Apart from the comments of long-term residents, who call this new Oakland Hills landscape "horrifying" (Schiewe, 2011), it is hard to realize that, before 1991, the Oakland Hills were a very different kind of space. Before the fire, the wooden houses were typically colonial style, and sat relatively small on their rather large lots. Vegetation was prolific, and tree canopies and luxuriant gardens gave the place a rural feeling.

Following Walker's typology (Walker, 1995), the pre-fire Berkeley and Oakland Hills were the materialization of the "ecotopian ideal": a community close to nature, drawn toward Libertarian political theory,

[5] To the panel of interviewees, the fire has had a particular meaning: some of those interviewed had lost their homes during the event, and many had relatives and friends that had been living in the Oakland Hills at the time of the fire.

and a bohemian lifestyle.[6] The first houses were built by an upper-middle class, and soon drew a population of artists and intellectuals. Because of its spectacular views of the San Francisco Bay and the Golden Gate Bridge, the hills became a residential choice for affluent families fleeing San Francisco after the 1906 earthquake. In the following decades, residents largely endorsed the ecological principles of John Muir, and were actively engaged in the preservation of great natural landscapes such as Yosemite Valley and the area's large sequoia trees.

The topology of the hills was scrupulously respected; lots and roads followed the natural perimeters of the creeks and woodlands. The houses reflected the residents' personal interests and philosophy: they favored a certain austerity in their architectural choices,[7] which was far from the eye-catching Victorian style of early San Francisco. Architects trained on the East Coast and Europe, such as Willis Polk, Bernard Maybeck, and Julia Morgan, were hired to craft the comfortable and—only seemingly (from the outside)—rustic cottages in an unassuming style that remains the Berkeley signature for houses not burned in the fire.

The tentative frame of the dimension of such event comes from different perspectives: from residents both outside and inside the fire zone, but also from news journalists, scientists and first responders agencies.

> All day long the television would show pictures, and the reporters were terrible. They would stand, and they said "We are on Martin Boulevard." Well … Martin Boulevard goes from miles away to very close [to our house], and they never told you the cross street! So, it was very difficult to get accurate information that day, but had the fire continued into Piedmont, that mostly would have been a disaster for me and my family. That was a frightening day. It was interesting because at the same time that I was here, in this house, my husband was driving into the real fire zone, with friends, evacuating some of their belongings, and so he kept telling me "it's far away, it's far away," but the helicopters would fly over here and say "Evacuate" … We did have a message to evacuate, so we sent our children with my mother to Walnut Creek, where we knew the children were safe, and we sent a few things, papers and things like that. But once my children, at that time, [they] were maybe 10 and 8, were safe, I felt a lot

[6] Together, these ideas and principles are probably what have earned the community the nick-name of "Nuts Hills," used by some.

[7] Some of these houses remain today, mostly visible in the Panoramic Hills, where residents have fiercely defended their space from encroachment.

> better about it. To know I only had to move myself and my husband. But that was close, that was the closest we've come, I think, to a disaster, was that day. [S.16]

The 1991 fire, like the Loma Pieta earthquakes was a collective event that, with nuances and in degrees, mobilized the entire Bay Area. The epicenter of the disaster was, of course, the Oakland Hills, where the blaze of the fire was the hottest. But invisible strings, or reticular connections, were streaming all around to create a new chain of actants.

For professional firefighters and disaster agencies, it was a "century event" (Keeley, 2005). Firefighter units came from Berkeley, Piedmont, Orinda, Moraga, the Marin and Contra Costa counties, San Francisco, and the California Department Forestry. Additionally, fire units, air tankers, and "spotter" aircraft came from as far away as Fresno and Salinas, responding to the alarm through the Alameda County Mutual Aid Coordinator, located in the Lawrence Livermore Laboratory.

During the fire, the destruction dispersed local families to friends and relatives further away, seeking refuge where it was available. Connections, even distant, even improbable, were activated.

> I had been in San Francisco the week before and met a graphic designer. And I called her up. I [was] thinking, "Oh my God, this fire is going to go straight to the Bay." So, I called up this woman that I had just met, and I said, "My hill is on fire. Can I come to your house?" And she said yes. Because she could see the stuff floating up. They had pieces of burning paper… she was over by Mount Parnassus by UCSF, and it was so windy, it was blowing over there. [S.16]

After the fire itself had been and gone, recovery efforts immediately went into effect—another moment when the borders of the disaster exceeded the strict perimeter of the disaster on the map. The Red Cross was one of the important actors in the recovery, but so were many neighborhood community organizations and churches all around the Bay Area. For the victims, these organizations brought an influx of donations, from socks and t-shirts to ovens and pots that "helped us survive in that first year. We weren't alone and we were helping each other," recalled a former Oakland Hills resident.

I found the traces of existence of these houses and the fire that destroyed them in a book, carefully stored in a closed, dark room, behind the huge concrete walls of the Berkeley Art Museum and Pacific Film Archive. Here, testimonies of the fire survivors were safely preserved from the ravages of time. These stories had been collected in a "Guest Book" during the exhibition "1991. The Oakland-Berkeley Fire Aftermath: Photographs by Richard Misrach", held at the Berkeley Art Museum and the Oakland Museum of California between 2010 and 2011.

The pictures, taken in the days following the fire, reflected the impact of the blaze on mundane and familiar objects. Their breathtaking views of the landscape that vanished under the ashes were so distressing that Misrach said he could not have shown the pictures before: "I made a radical decision for an artist—to put my work away. But this was in my backyard; it was a major trauma, and people died" (Taylor, 2011). Thus, a part of the history of the Bay Area had been hidden for 20 years. Because of this, the exhibition itself was a different way to experience the disaster, mixing art with community building, oral history, with individual and collective catharsis: "This exhibition is like no other that I have had [...]. I have been moved and stretched emotionally and intellectually. Every new story I hear from people is a little revelation" (Taylor, 2011), confessed the photographer. For many former Oakland residents, the exhibition worked as a relief, a long-awaited public acknowledgement of the dramatic dimensions of the fire. When the exhibition concluded at both venues, the Guest Book was donated to the Berkeley Museum to be kept in the collection. "Just as with the Civil War memories, this is a timepiece," commented the photographer (Taylor, 2011).

3.1.1.1 Living Through the Fire: The Emotional Measure of Space

This is where I found it: a piece of both art and history in a museum collection. Accessing the book, where the testimonies had been consigned, was like getting access to a well-guarded secret. After entering the museum, the person in charge took me to the storage room. The room had no windows and was closed in by two fire-proof doors. Once inside, the dark room looked like a long corridor. The museum collections were stacked on wooden shelves: entirely wrapped statuary, paintings, and curios. The book was waiting for us, thoroughly enveloped in several layers of silky paper and a waterproof piece of plastic. Before removing the last layer, my guide put on surgical gloves, handing another pair to me and, only then, carefully displayed the large black book on the table.

It had the color of ash and the title, "1991," was embossed on the front cover. Inside, the drawings and comments of visitors reflected on the multiple perspectives and vantage points from which the Oakland Fire was experienced. "Now I remember, it was a Sunday," recalled one of the anonymous visitors. Another commented on the spiritual loss that the photographer has managed to capture—the presence of the absences, the presence of grief and loss, and the ephemeral aspects of life. Thanks to the power of the pictures and the comments found on the pages of the book, the space of the disaster, the border of the controversies, changed. Pictures of the devastation did not talk so much about the value of the losses, but the loss itself. They focused on the space left empty after objects were gone: a house, a plastic toy, a life. Many testimonies detailed the experiences of the fire, the spectacle of the unstoppable devastation. They brought back testimonies from elsewhere, mixing time and place, references and events, spreading the space of "disaster" further to other past catastrophes. The exhibition, the books, and the comments responded to each other and connected people with distant past experiences, building bridges with each other's stories, times, spaces, and fears and trauma. They shared feelings that the pictures of devastations left open passages for the recounting of personal history, poetry, and confession. The fire, the experiences it yielded, and its display, reminded people of ancient myths, such as the one about the Indian Godess Sita who perished in flames. The experiences of the proximity to death, of course, was often mentioned, as well as the more metaphysical questions that the pictures tackled.

The experience of the fire is a collective one that goes beyond the limits of this work but, additionally, the collective experience moves beyond a narrow understanding of the events of that day. It tells the story of a dispersed "tribe," as an anthropologist would say: it is a story that brings together generations, morals, totems, values, beliefs, celebrations, and a wide range of human emotions including fear, sorrow, remorse, surprise, sadness, curiosity, excitement, and so on. This tribe is like others that form our modern world, one that recognizes itself for the duration of an event and that had to find a collective answer to the questions of what really matters, and what should be taken into consideration. Where should we draw the line between what is inside and what is outside our tribe?

As opposed to the singular form that is used to describe it today—"The" Oakland Hills Fire—there were many fires, many fronts, many actions, and many actants involved. The FEMA report produced after the dust had settled (Routley, 1991) established multiple correlations between climatic conditions, vegetation, land development, narrow streets, and previous occurrences of fire and regulatory precautions.[8] The report documents the initial, incendiary moment of origin as Saturday, October 19, but then it was rekindled on Sunday, October 20. The report provides written descriptions of the complex layering of the space delimited by multiple boundaries of different types. For the firefighters, the actual, physical terrain was difficult, dry, and steep, with "boxes" and paths which the fire could engulf to become stronger:

> The terrain [...] rises abruptly to form a row of hills called the East Bay Hills or the Oakland Hills, with a ridge line approximately 1,300 feet above sea level. The ridge line runs generally in a north-south direction, parallel to the shoreline of San Francisco Bay and approximately five miles inland. The hills separate the coastal flatlands from the inland valleys of Contra Costa County, and the ridge line establishes both the eastern city limits of the city of Oakland and the eastern boundary of Alameda County. (Routley, 1991: 5)

The fire was at the interface of several ecosystems that had been transformed over the years, such as by droughts in previous years, as well as by the importation of invasive, non-native plants such as the very flammable eucalyptus tree. The economic dimension played a role: progressive urban development of the hills and the design of the expensive properties did not allow an efficient fire-protection wall:

> The homes on the hillside range in value from $250,000 to several million dollars. The steep slopes of the canyon walls present difficult construction challenges. Many were built on platforms overhanging the canyon walls or with multiple levels stepping down the hillside. Garages, sundecks, or swimming pools were often constructed on the top level, with two or

[8]The FEMA report was realized by a private consulting firm, Tridata Corporation, which, according to its website, "is one of the nation's leading public safety consulting firms. We specialize in research, analyses, and management studies in fire protection and emergency medical services, prevention and preparedness, and homeland security. Our array of fire and emergency research extends across the U.S., supporting the federal government; state, municipal, and local governments; and the private sector" (http://www.sysplan.com/capabilities/fire_ems/index.html).

> more levels of living area below the level of access from the street. Short bridges were required in many cases to span between the street and the garage entrance. Untreated wood shake or wood shingle roofs were common, and no requirements for fire resistive roof coverings or walls were enforced. (Routley, 1991: 7)

In addition, residents' beloved eclectic vegetation was so dense that it often covered roads and electrical lines. These combined factors resulted in an inferno that led the authors of the FEMA report to conclude:

> the fire was beyond the capabilities of the fire suppression forces to control. The stage was set by a number of contributing factors that created the opportunity for disaster. When the Santa Ana wind condition was added to those risk factors, the combination was more than any fire department could handle. (Routley, 1991: 45)

When the firefighters were called that Saturday afternoon, nothing seemed to be really worrisome. The small fires were under control, and crews left at night, confident that this was a minor event. But on Sunday morning, matters took a very different turn:

> The sudden eruption occurred when a firefighter was digging out a hotspot, near the perimeter, and sparks were carried into an area of dry brush, which virtually exploded into flames. The sudden development of the new fire was witnessed by several individuals, including experienced firefighters, who were able to describe the phenomenal rate of fire growth and spread. (Routley, 1991: 48)

The exact timeline during the couple of hours that followed the brutal eruption of the fire is difficult to establish. Actions became reactions to follow the specific deployment of the fire. The FEMA report describes it as follows:

> The synergistic effects of the wind and the thermal energy released by the fire created unusual fire phenomena that exhibited some of the characteristics of a fire storm on a localized basis, but the term conflagration is more appropriate for the overall situation. (Routley, 1991: 48)

In the first hours of the fire, the large combination of factors described above created a worst-case scenario for the wildlands and urban interface.

Communications between firefighter units were interrupted, the initiated mutual aids between Bay Area Fire Departments fell short,[9] lives were lost but some were saved; houses burned while people jumped into their pools to save themselves while the fire jumped across the freeway. Finally, the historic Claremont Hotel was saved, as well as the rest of the East Bay, because at around 5 o'clock, the wind fortunately stopped blowing. For more than six hours, firefighters were overwhelmed, and anybody who could help did so. The boundaries between safety and danger first responders and the victims were blurred as the fire transformed reality as they knew it.

The fire was a phenomenon that, with no warning, brought multiple and, until then, dormant chains of actants together in a spectacular movement that surpassed the response capabilities of the firefighters. The wind, the canopy, the decks, the drought, the steeps hills, and all else became a gigantic blaze. Flames could be seen for miles around and the giant cloud forming above the hills was a major source of preoccupation for inhabitants of the Bay Area. From the streets of Oakland to the hills of El Cerrito and San Francisco, everyone was looking toward the fire, toward the phenomenal event that imposed its striking presence.

Never in the history of the USA had people had to face an urban fire of this dimension (Hoffman, 1998). In the course of one day, the losses and the transformations were tremendous. If some people made it out of the fire, if some houses and some lots were saved, others were not. Assemblages were transformed and categories mixed.

The best way to give life to the complexity of the event once more seems to be through the survivors' own words, the literary translation of the survivors' experiences. For residents who were in direct contact with it, the fire was not a scientific event that can be described with objectivity. It was an imposing, breath-taking vision, inspiring distrust and fear, forcing people to make rapid decisions involving their lives and the lives of those around them. The slow pace of the survivors' storytelling allows listeners to pay attention to the diversity of perspectives, and the thickness of those narrated experiences. In these stories, a world absent

[9] Firefighting units from different counties were unable to use Oakland hydrants because of the differences between their hoses and the sizes of the hydrant outlets.

in scientific scenarios but present, in its way, in fictional ones, emerges[10]: the collision between the ordinary and the extraordinary, the transformation of space and subjects.

These fire survivor narratives express the emotional dimension of going through the disaster. The survivors tell us about the destruction of their worlds, sharing with readers or listeners this sense of loss, chaos, and confusion. Starting with a written, perhaps more distant, account, the following quotations strongly evoke the fire's velocity and its consequences on a once familiar environment.

In the first excerpt, a survivor wrote a letter to his spouse describing the intensity of the situation and the context in which their house was partially burned. In 2012, a couple of years after her husband's death, the addressee integrated the text as a narrative in a documentary she produced for the commemoration of the twentieth anniversary of the fire. In the second excerpt, drawn from a novel, we find the narrator away from home when the fire starts; she is driving back to save an almost finished manuscript and some personal items. The third example, an excerpt from an interview, tells the story of the day that her house burned down and how she escaped the fire.

> Early that morning, out of the East to the back of the house, I saw a rail of smoke up of the hills, several canyons away and a few hundred feet higher than our house. [...] Blue sky and dry air suddenly turned to burning orange light, hot sun overhead broken by waves of furious swirling hot winds, loaded with flames. Darkness at noon. The air, a blast furnace. An inferno. Terrifying memories of fires burns in memories of those who live in Northern California. [...]. One house ignites every eleven seconds and explodes. [...] Winds and flames blown in all directions, creating destructive chaos, people abandoned their cars to escape on foot as the fire engulfs the hillside neighborhoods. [...] Out through the back of the house, I could see trees in flames a few hundred yards away down in the field below me. And our roof was burning. At least twenty neighbors appeared to help with fire hoses. The firemen did not even try to evacuate us. They needed help. But at three o'clock they evacuated. One of them told me that saving our house, had saved the neighborhood. (Mitchell, 2011)

[10] In an interesting move, some of the quotations have a status between several modes or types of narrative: they are journalistic accounts rewritten from personal correspondence or diaries, or they are in the form of biographic novels. In each case, the literary forms are the necessary condition for the transmission of what matters to the authors: the transmission of an emotion and an experience.

With their subjective dimensions, these accounts of the fire tell us about the transformation of the environment and, often, of the narrator. Everyday movements are changed and organized around one predominant fact: the swirling fire, pushed by the Santa Ana wind. The movement of bodies was springing into action by the urgency of the moment; as well as the movement of 'things' being transformed—mostly into ashes—and the movements of emotions and thoughts. Space was shrunk or extended, as was time: the space of the fire that day was difficult to grasp. In the second excerpt, from a biographical, but fictional book, the narrator is away from home when the fire starts. As she has no precise information about it; she is trying to make sense of the disparate, unclear, and sometime contradictory information that she receives through the radio, hearing the disaster in the act of "becoming":

> I was driving home from funeral ceremonies for my father. [...] I turn on public radio for the intelligent voices, and heard that the hills were burning, toward Moraga, toward Walnut Creek. It's not my poor sense of direction, I told myself, but the newscasters in confusion. The perimeters of the fire were different from station to station, from tapped news, to live news. North of the Caldecott tunnel, south of the Caldecott tunnel, east, west of the Warren Freeway. I picture wildfire far up in the hills – ridgelines of flame spilling down, then running up sere-grass slopes. [...] Impossible that it crosses ten lanes of freeway and take over the settled, established city. [...] Here the winds and all seemed normal; I had no evidence that hurricanes of fire were storming on the other side of the hills but for the radio. "Forty-five houses have gone up to flames." "About a hundred homes." "A hundred and fifty structures have burned." The number would keep going up – nine hundred degrees, the temperature of molten lava; twenty-one hundred degrees, the temperature of kilns, thirty-five hundred houses. "Wind of forty-five miles per hour..." "...sixty-five miles per hour fire wind..." "...record heat and winds" ... "foehn wind." "North East wind [...]" "The fire has jumped the junction of highway twenty-four and highway thirteen." "It's blown over and through ten lanes". [...] "If the Claremont Hotel goes, explodes, the fire will burn to the Bay." (Kingston, 2003: 4–5)

For many residents, adjusting to this changing space became a survival skill. People had to develop special spatial competency both with "the space as they know it," but also with the space-in-transformation. The third testimony is from inside the fire zone. This time, the narrator is dealing with the need to escape, the urgency and the difficulty in making good decisions and finding good ways through danger zones.

> I have twins, well they had a little birthday party down in Berkeley and I remember taking them to that birthday party and seeing the smoke and thinking what would I take with me if I had to evacuate? And then that [the first fire] got taken care of. And the next morning they had another party to go to and I'm sewing and taking them to their party, and it was early, it was like 9:30 that I had taken them to this party and on the way back I had seen smoke again, and so I called the fire department and I said you know, there's still smoke in the same place. And they're like, "No, no we have it, it's ok." And so, I'm like, "alright." My husband's out here gardening, it's getting very windy, and I'm sitting here sewing something for my older daughter. It's getting darker and darker outside, and I'm getting very nervous. My daughter had a friend who lived at the top of the hill. So, I called them up and I said, "Do you know what's happening?" and they said, "No no, I'll find out." So, he's going out to look. And in about 15 minute later, and this is about 10:00/10:15, my neighbor from across the street calls me and says, "What's going on, they closed the tunnel?" He had been playing tennis down by the Country Club, the Claremont. And I said, "I don't know but I asked a friend at the top of the hill and I'm waiting to hear." Now this guy says, "I'm not waiting; I'm going to find out." So, I'm a little worried and I call, we have a surrogate-grandmother type who lived up the hill, and I was worried about her, and I called her up and I said, "Are you okay?" And her daughter picked up and she said, "Well, we're leaving. We're taking her." And I said, "Okay, that's good." Then immediately I get a phone call from my friend up at the top of the hill. It's black smoke up here, oh… okay. He hangs up. And then I get a phone call from Warren across the street, "It's crested the hill. Get out now." And I'm thinking, "Oh, it's coming from the top of the hill." Well, I'm running out to get my husband […] I'm yelling, "Gordon, Gordon, Gordon!" He's hard of hearing and, you know, the wind is blowing, you can't hear, and he comes running and I say, "Warren says it's crested the hill," and he says, "Yes, I saw it, let's go." So, he goes in one car and takes the dog we had before this and I tell my daughter, "We need to go." […] And as we backed out of the garage, I remember closing the garage door thinking, "Well, that way robbers won't come in," and I'm backing down my driveway and there is a line of cars going bumper to bumper down the hill. And one of my neighbors smiled and let me in, and we're going down the hill and I see a friend of ours at the top of the hill waiting, and we're saying, "Get in the car!" and they're saying, "No, no, no, we're waiting for my husband, he's gonna come!" When

you leave here you'll see that there's a light and it's this funny little intersection. Well it's bumper to bumper. And here is this place that is now a garden but wasn't a garden, and the hill to the other side has a lot of eucalyptus trees. And the wind is blowing and these fire brands are blowing over us and we're stuck at the light because it's bumper to bumper traffic. And I edge up to the light and across the way, here's Highway 13, here's tunnel road, here's the way to go to 24, and here I am. Well on the downhill slope is a grove of eucalyptus. As I'm stuck here, it's like being in the movies, it just went "VOOM!" It just burst into flames. And I'm stuck here and it's so hot because it's burning down the hill, I'm in my car with air conditioning, sitting, leaning to the right because I can't sit straight it's just sooooo hot. [The fire] is coming down here. It was so hot I couldn't sit straight. And I'm thinking, "I gotta get out of here. I gotta save my daughter." And here are the flames, and we turn the corner and all of a sudden, we see people running. What had happened was by the time I turned the corner; the fire had crossed the road and people couldn't get on. So, we turned the corner, still bumper to bumper going down the road to Tunnel Road, which you know is narrow, and it's divided, and I'm thinking, "I can't do this, this is taking forever." Well, there is a place, a cut in the divided road and I turned and decided, "Well, I'll get off the road." And I'm going down this road called Roble Road, really narrow, and then I'm thinking, "Oh God, I'm going right to where the flames are." And I can't turn around, people are following me, and I'm just praying "Please, please, let me get there." So, I go down the road and it lets you out near Chabot Elementary School. It's down the hill. And now this is really weird. It's like a curtain. Black sky, blue sky. Not filtering just big like a wall. Black sky, blue sky. People standing out by Chabot School with their beer looking at the fire coming in. We get there and we go, "Yay, we're safe." Go around Chabot and go under the [Highway] 24 to go up Broadway, that's the first policeman I see directing traffic. And I just said to my daughter, "We're gonna go pick up the kids." So, we go to Piedmont, which is where they were, rushed out of the car, they must've thought I was a crazy lady. I was knocking on the door saying, "You don't know but there's a big fire, I probably lost my house by now." They had no idea. Get the kids in the car and I think, "Where am I gonna go?" So, I go to Alta Bates Hospital, which is down in Berkeley and where I used to work. And I said, "At least there, there are phones, there's food." [I] bring the kids in, and my daughter brings her bag with all her stuff in it, and one of the things that I took was my little book, address book. [S.16]

These stories help us to understand the transformation that the fire imposed on that space. Distance, identity, function: everything is moved,[11] and can only be defined when things stop changing. For those in the direct grasp of the fire, it was an important moment for the hybridization of their identities. Suddenly, in the space of a couple of hours, the people directly involved had to adapt to completely new situations and new spaces; they had to become something that they had not been before, and never imagined themselves to be: a victim of a disaster, a rescuer, a volunteer firefighter, a runaway, a survivor, a member of the homeless, a refugee, a community organizer, an orphan, a widow or widower, or a childless parent. The transformation impacted all, without consideration for previous roles or functions. The fire did not discriminate. He, who in a previous moment would have been an expert in a particular role—a policy analyst, an engineer, a policeman, a firefighter—was now running for his life. She, who previously belonged to the ordinary, was now fighting to save a school, or a house. The roles, like the boundaries, became mixed; the lines became blurry: knowledge, power, values—all of these well-established actants, or so they thought—flew into pieces and fell to the ground in a shape no one could recognize.

3.1.1.2 Transformation of Space After the Disaster

During the Oakland Fire, the well-known familiar living environment was reduced to ashes. A land of desolation now stood where a lively neighborhood had once been. It left the residents looking for meaning. The chains of signifier and signified were disconnected, interrupted with other images and references. "You've seen those pictures of bombed-out London after the war, that's what it looked like," [S.16] explained a respondent. In her memoirs about the fire, *The Fifth Book of Peace*, writer Maxim Kingston recalled her discovery of this new environment:

> I came out into a changed world. Its color had gone out. Its dimensions had stretched away here, shrunk there. New mountain and canyon vistas as far as I could see. [...] What I wanted to see, what used to be, popped in and out of sight, alternated with the real. The hot ground was reeking mirages that cheated the eye with blear illusions. A thing would appear—a chimney, an oldened wrought-iron gate, a ceramic pot—but it did not

[11] "It was a windy, hot, dry day, and 2,500 houses went up, just 'poof.' [...] And then there was the rest of day, where it's kind of moved slowly, but gradually. Because the wind died down. And it wasn't until the wind died down that they were able to put the perimeter around it, and stop it" (S.14, 2009).

> cue the next thing, the thing that should be attached to it (house, fence), to appear. Things were out of the order that was in my mind. Memory was off. If only I had paid better attention – I have to be more awake – I would not be losing the detailed world. (Kingston, 2003: 10)

Loss is what remained long after the last burst of flames was extinguished. Years later, at the time when I was conducting these interviews, people were still experiencing the loss: looking for things that had burned 20 years ago, sometime confusing what belonged to the past and what exists in the present: "I found myself going to closets that don't exist anymore and remembering if we still had something. The first few years you kind of forget: 'Oh right, we don't have a big turkey pan anymore! The gravy bowl, too…,' things like that," explained a survivor. For the people who lost their homes, they lost an important part of their lives. A world of ordinary, used, transmitted, familiar, polished, and loved objects, which disappeared in a day. The multitude of the mundane, reliable tools that carried them through their everyday lives had vanished. Years of patient work molding houses and gardens, and events and routines, had gone up in smoke. The destruction of cherished pictures made one survivor comment that "even the presence of the absence" [R.5] was missing. For many, the fire ruined years of work and ideas. Photographers lost their entire archives, authors lost their manuscripts, architects lost their plans, analysts lost their notes.

> In the fire I lost my home and my possessions. I lost my clothing, furniture, photographs, heirlooms, artwork, beloved objects, one car, and two pets. Since my office was in my home, I also lost twenty-five years of anthropological research, seven manuscripts not yet into publishers, all my other writings, ideas, projects in development, the slides and photos of travels, lectures and courses and notes, and my entire library. (Hoffman, 1998: 55)

The fire destroyed the material connection between generations, between residents and the world, and between residents and themselves. For a time, it reduced to zero the possibility of communication and transmission, and dragged the survivors into a situation of distress that, even years later, was hard to explain: "Describing the devastation both physical and psychological of this kind of loss is like trying to define eternity or infinity. It defies words, evades phrase, and renders mute any and every euphemistic catch all" (Hoffman, 1998: 55), recalled a survivor. To deal with this immense grief, people called up ghosts, of a sort: "Even when I was really upset about losing my mom's jewelry and all the things she gave me, I heard her voice in my head telling me: 'it's just stuff,'" [G.6] said another survivor whose mother had passed away just months before she lost all of her belongings in the fire. For others, the particular nature of this experience, the bitter emotion that it generated, found a way out through the expression of long-forgotten languages of their ancestors. A survivor recalled that after the fire, her Chinese-speaking mother would tell her, "Don't *hun* things.": in Cantonese "*hun*" is the word for pain and loss (Kingston, 2003). What people soon realized was that with the loss of these objects, the signposts of their previous existences, they lost much of their entire worlds.

> We were integrated in a way that I have not seen in other places: rich people and not-so-rich people lived side by side. Some of us were owners, some were renters [...] we had blind-chosen one another at random, and were not trying for it, but we'd gathered a community. It is over. No more chances to improve on and appreciate what we were. Too late. I never joined the Christmas caroling organized by peaceful Christina next door. I hadn't gone to any Randy and Sue's Halloween open houses. They went door to door inviting one and all to come trick-or-treat and party. (Kingston, 2003: 18)

Thus, more deeply, the fire dismantled the foundation of the social and cultural organization of the community. While losing their houses, women survivors also lost their capacities to interact in the public sphere as equals to men. Confronted with the Herculean task of rebuilding the subtle compositions of their existences, as well as those of each member of their families, many women quit their jobs, and spent a considerable amount of time collecting, storing, and sorting objects necessary for their future everyday lives, while men most often dealt with the public aspects of the reconstruction: insurance companies, architects, engineers, and so forth. In a community that had long opted for shared domestic responsibilities, the disaster put many women back into disparaging forms of social organization, gender roles, and values, thus fundamentally transforming them and their families.

> The women of the community were independent, men equitable, couples by and large egalitarian. People of [the] both genders occupied the same

> segment of space, public and private arenas, hours of day and night. But for many, progress in carving out a new gender behavior suffered a fifty-years setback. In the shock of loss both men and women retreated into traditional culture; realms and personas. (Hoffman, 1998: 57)

As Hoffman noted, "The return of kinship became, as it had customarily been in our traditional society, women's jobs to facilitate" (Hoffman, 1998: 58). Women became the connectors within their families and, in the process, lost their social alliances. Here, some chains of action were reconstituted, while others were broken altogether.[12]

> Every survivor suffered the wrenching shifts in former associations. Friends did not, or could not, offer aid or comfort. Friends grew impatient, proved unsympathetic, disappeared. What did maintain for most were the links that lie more deeply rooted in our society, blood kinship ties. (Hoffman, 1998: 58)

As a consequence, the drifting Oakland survivors found themselves more and more excluded from the rest of the Bay Area. The mixed emotions brought up by their situation secluded them from the rest.

> At first the outside community saw us with sympathy. Eventually, when recovery took longer than the day, the weeks or months they envisioned, they came to view us as greedy whiners and underserving receivers of pots of gold. [...] As the community separated itself and jealousy erupted, it was often once more the women who bore the brunt of it. [...] Targets of wanton envy, we were informed that we had all new things, we would eventually have new houses, and we were "lucky." (Hoffman, 1998: 60)

The space of the disaster is a space of loss.[13] It is marked by the web of relations that connects—or have stopped connecting—absents and

[12] "Women in our society, as part of our domesticity, act as social connectors. We are, to a large extend, the linesmen of our ties and the 'bondsmen' of our everyday social circles" (Hoffman, 1998: 59).

[13] The term "space of risk" was coined by Michael Taussig in an influential and provocative essay that tackled the culture of terror created by the colonial exploitation of rubbers trees in the Putumayo Region in Peru. In the context of Taussig's study, "the space of death is here the Indians, African and white gave birth to the new world" (Taussig, 2009: 5). In Taussig's research, the space of death is defined as a web where European and Indian understandings of evil and the underworld, their cultural backgrounds came into contact

the new-comers. This is a space where the modes of existence change substance, a space of transformation. It is a space that pushes the capacity of individuals to their limits to think and evolve in a world that they thought they had known. This space is transitory, it is the "in-between," a life lived between the disaster and reconstruction, in-between what was and what will be.

The survivors found themselves in the process of grieving and trying to recover from the disappearance of their homes and their neighborhoods, and the deaths of 25 people. In its destruction, the fire had engaged a displacement of priorities and needs. Every single "thing" had changed: "The patterns of my days, my plans, my routines were irrevocably ruptured. The warp of my past was torn from the weave of my future. Who I am, what I was, what I intended to do, the fabric of my life, utterly unraveled" (Hoffman, 1998: 56). With this transformation of the space, large perspectives were changed, systems of relations were reoriented, and even the awareness of the experience was questioned:

> To be part of a major natural disaster is a humbling experience. As I live the aftermath of the disaster, I find myself in awe of nature, as another human being woven into the web of life on earth. It is the feeling, I think, that binds us together to support one another and to listen to one another's stories. We were not really different from the medieval peasants in the stable hearing the saga of their clan over and over again. Or the veterans of a war, reminding each other of their miraculous survival. (Adler, 1992: 58)

After the disaster came the time of reconstruction: the recomposition of a neighborhood, which slowly again became reintegrated into the surrounding space. After such a loss, reconstruction was, of course, not easy. The next quotation gives an excellent example of the complex range of emotions that one could attach to their burnt properties.

> We didn't rebuild, we bought. We rented in Orinda after the fire and for a long time we didn't know what to do. It was after we were married but

with one another, and metamorphosed each other: "The space of death is preeminently a space of transformation: through the experience of coming close to death there well may be a more vivid sense of life; through fear there can come not only a growth itself consciousness but also fragmentation, then loss of self-comforting authority" (Taussig, 2009: 7).

before children, but we knew children were in our future and where we lived was not a great place for kids, high in the hills, small house, small windy roads, not great schools, so we always knew we'd have to add-on if there was room, move, go to private schools. So, when we rented in Orinda we got to know that community: it had excellent schools and it made more sense for the next phase of our lives. That became an easy decision for us, but then we hung onto the property, maybe because there was an emotional attachment and we didn't know if we were going to do a spec development, build, and then sell it. We couldn't even deal with it because there was still an emotional attachment, so it took a number of years before we had enough of a distance to say, "Yeah, just let it go." But it was still really emotional, selling the land. It made me realize: the economy was such that people started contacting us about selling the land. It was a small, but really special site, and I saw a lot of building going on in the hills I didn't like, building big boxes to the limits. So, given that the site was special and the [career] field I was in, I didn't want to just sell it to some spec. developer who didn't really care. We ended up selling it to someone who was an architect-builder and we ended up working together, the realtor understood my concerns and set up a meeting between me and the architect and they were really reassuring me of how they approached design and how they cared about the site and it made me feel better emotionally about selling the land. That was really important. I remember I would drive by the land a lot, watching it being built. For a long time, I'd go to the site and we'd kind of walk around and stand there and feel things. Then I started getting numb as my life moved on, I'd drive by the site and there wasn't a lot of emotion. I'd watch them building the house and there wasn't a lot of emotion. One day I drove by and the house was almost complete and somebody saw me outside, one of the builders. He said would you like to come in and see it, and I remember sure, at the moment there was no emotion, so I walked into the house and it was so beautiful, and I just started crying. I realized it was because it wasn't a crying of sadness, it was a crying of happiness. The woman must have thought I was wacko, because she had not lived in the Bay Area, so she didn't really get it: this crazy woman walks into the house and just starts crying and saying over and over again, "I'm so happy for you. It's beautiful and I'm happy for you." It was like closure, it was truly like happiness that somebody took this land and did something really nice with it and it was beautiful and it healed and it was like closure. (She starts crying.) It completely surprised me, as I'm surprised right now. I've been pretty bland this whole interview, it's like I'm so distant from this experience, I'm surprising myself that describing this moment can still bring this out. I'm completely shocked by why I'm crying right now, but it must say something… [G.6]

At this point, the interviewee asked me to turn off the recording device, leaving open the possibility of interpretations. However, the story of G. shows how integrated the connections between space, emotion, and building space can be.

3.1.2 A Science and Technology Tale from 1906

Coming from Stanford, where he was staying at the time of the San Francisco earthquake in 1906, James, the pragmatic philosopher, showed great interest in the complexity of the disaster both as an object of science and as a subjective phenomenon. In an essay called "On the Mental Effect of the Earthquake," James discussed the articulations between scientific and lay understandings of the earthquake, as well as his own comprehension of the "existence" of the phenomenon. As we have noted in the Introduction, James details some of the existence of the earthquake:

> For "science," when the tensions in the earth's crust reach the breaking point, and strata fall into an altered equilibrium, earthquake is simply the collective name of all the cracks and shakings and disturbances that happen. They are the earthquakes. But for me the earthquake was the cause of the disturbances, and the perception of it as a living agent was irresistible. It had an overpowering dramatic convincingness. (James, 1906)

James was aware of the articulations needed to create a chain of actants defining, for scientists, residents and himself, the earthquake. The 1906 Big One was, at the same time, the "collective name of all the cracks and shakings and disturbances that happen," the "living agent" and the fear, dismissed, of the "final judgment." During his two trips into the city of San Francisco, just after the disaster, James was surprised by the absence of panic, pathos, and anguish; further, he was very impressed by the residents' capacity to become organized through:

> order makers, whether amateurs or officials, [who] came to the front immediately. There seemed to be no possibility which there was not someone [*sic*] there to think of, or which within twenty-four hours was not in some way provided for. (James, 1906)

He made the point that understanding of the phenomenon cannot be only a rational experience:

> I realize now better than ever how inevitable were men's earlier mythologic versions of such catastrophes, and how artificial and against the grain of our spontaneous perceiving are the later habits into which science educates us. It was simply impossible for untutored men to take earthquakes into their minds as anything but supernatural warnings or retributions. (James, 1906)

Making these observations, James was really preoccupied with "subjective phenomena exclusively" (James, 1906), and his desire to decipher the nature of agency from the various perspectives of his own consciousness, from science, and from the collective reaction to the earthquake and his reflections invites us to move a step forward in the possibility of taking seriously the density and intensity of what is given by the experience of earthquake. After the San Francisco earthquake, James returned to the East Coast and worked on the concept of "radical empiricism," which Bruno Latour later translated as: "we don't want more than what is given in experience, [...] but we certainly don't want less either" (Latour, 2005). As the British geographer Nigel Thrift noted, this approach is thus very interested in actual, physical practices[14]:

> Practices are productive concatenations that have been constructed out of all manner of resources and which provide the basic intelligibility of the world: they are not therefore the properties of actors but of the practices themselves. (Amin & Thrift, 2007: 8)

3.1.2.1 The Legacy of the Big One

If a consensus around the definition of Oakland's "event" has not been yet been reached, the same, in fact, is true for the 1906 San Francisco earthquake. The real and full impact of the 1906 earthquake and fire, which ravaged San Francisco, has been difficult to appreciate for decades: not only because analyses of the consequences of this earthquake, and the fire that followed, have been continually mixed together, or because survivors' and eye-witnesses' testimonies were long-neglected, but also, because the earthquake did not fit in with the San Francisco founders'

[14] "In 1934, geographer Henri Lefebvre noted: 'Upon the basis of acts repeated billions of times (practical, technical and social acts, like the acts of buying and selling today), customs, ideological interpretations, cultures and lifestyles erect themselves. The materialist analysis of these styles has progressed very little'." (Lefebvre, 1934: 72, as quoted by Thrift, 2008: 147).

idea of the city. Recent research, conducted in archives as well as in earth sciences laboratories, has permitted a reassessment of the real dimensions of this event within its true historical context (Hansen, 1989; Tobriner, 2006) (Fig. 3.1).

If the event is hard to define today, it is because it was hard to believe then. The unfolding succession of events—first, the earthquake; then, the fire; the decision taken to dynamite standing buildings to avoid the propagation of fire; the use of the National Guard, and so forth—have created a narrative that has taken liberties with what was really happening. The above image is, in this respect, an interesting story inside the story.

> There is particularly one scene that I recorded the morning of the first day of the fire (on Sacramento Street, looking toward the Bay) which shows, in a pictorial effective composition, the results of the earthquake, the beginning of the fire and the attitude of people. On the right is a house, the front of which has collapsed in the streets, the occupants are sitting on chairs calmly watching the approach of the fire. Groups of people are standing in the street, motionless, gazing at the cloud of smoke. When the

Fig. 3.1 Looking down Sacramento Street San Francisco, April 18, 1906 (*Source* US Geological Survey Department of the Interior/USGS/U.S. Geological Survey/photo by Arnold Genthe)

> fire crept close, they would just move up a block. It is hard to believe that such a scene actually occurred in the way the photograph represents it. (Genthes, 1936)

The first aspect of the controversy focuses on the qualification and the definition of the event, which was intimately connected to the challenges and the intellectual framework of the time. Questioned on the topic, one respondent noted, "1906 was a mega-catastrophe but there was so much denial on the part of folks" [J.8]. Denials of the event were indeed coming from a variety people and sometime for contradictory reasons, which shows the difficulty of assessing the multiple risks from different vantage points.

> Early historians chose to emphasize the fire over the earthquake, and in the decades afterward, 1906 was remembered for "the Great Fire." Some recent historians charge that this was a cover-up, geared toward reassuring investors that San Francisco was not a peculiarly disaster prone place, since fires, unlike earthquakes, can happen anywhere. (Solnit, 2009: 41)

Other discussions have concerned the actual state of the destruction caused by the earthquake. Former sources claimed that more than 28,000 buildings were destroyed by this earthquake,[15] while more recent

[15]"Just before 5:12 a.m. on April 18, 1906, a magnitude 4.0 foreshock on the San Andreas Fault quietly rumbled throughout the Bay Area. About 20 seconds later, a magnitude 7.8 to 7.9 temblor began to rupture, with its epicenter below the Pacific Ocean, just 3 kilometers west of Ocean Beach, San Francisco. Violent shaking swept throughout the entire region and included 17 serious aftershocks within 1 hour. The quake ruptured 477 kilometers of the San Andreas Fault between San Juan Bautista to the south and Cape Mendocino to the north. [By comparison, the Bay Area's 1989 Loma Prieta earthquake had a rupture length of only about 35 kilometers and one-thirtieth the energy.] The rupture propagated up to 5 kilometers per second. The amount of horizontal displacement between the Pacific and North American plates varied from 0.5 to 9.7 meters. The earthquake was felt from southern Oregon to Los Angeles and inland as far East as central Nevada. Kevin Starr, professor of history at the University of Southern California and California State librarian emeritus, described the earthquake during a presentation to Lawrence Livermore employees last April. He termed the quake 'one of the greatest catastrophes in U.S. urban areas.' In all, more than 28,000 buildings were destroyed, many of them unreinforced structures that collapsed instantly. From a population of about 400,000, the earthquake killed approximately 3,000 people and left 225,000 homeless" (Heller, 2006).

sources tend to show that the earthquake did little damage when compared to the fire that followed.[16]

> The major buildings of downtown San Francisco [...] appear [...] undamaged. Even buildings in construction appeared unhurt. The earthquake resistant buildings of the 1870s, like the Palace Hotel, the Occidental Hotel and the Appraiser's Building, appear intact. [...] The overall impression from our vantage point is that most buildings, including the overwhelming majority of brick buildings, survived the earthquake with little exterior damage. (Tobriner, 2006: 119)

Of course, changing the reading of the event changes the conclusions and recommendations for future events. If it was the fire that brought down the city, and not the earthquake, what does this mean about the earthquake readiness of San Francisco's buildings? Should we, following Stephen Tobriner, conclude that the 1906 earthquake "has proved the capability of our structures to withstand such vibrations of the earth as are liable to occur in the future" (Tobriner, 2006: 50)?

The 1906 earthquake is a phenomenon that never found its total existence, its own story, because the reading of the earthquake event was overlapped by the fire that occurred in the earthquake's aftermath, as well as the wide number of experiences, stories, and interests interwoven at the time. Time and distance have washed out the overwhelming effects of the disaster; and the survivors,[17] who were young children at the time, are no longer here today. If the history of the 1906 earthquake has been marked by a denial of the materiality of the space, the awareness of the specificity has not been totally forgotten.[18]

[16] Looking at pictures taken directly after the earthquake but before the fire, Tobriner has established a new estimation of the 1906 destruction. In doing so, he has also contested what "many scholar and popular historians have accepted and repeated [namely] the idea that San Francisco of the 1860s denied the existence of seismic danger [...]. However, historical records show that architects, engineers, and even everyday citizens understood the consequences of the earthquake of the 1860s and tried to inventory the damages, to understand what had happened, retrofit building to resist future earthquake and to build earthquake resistant structures" (Tobriner, 2006: 35).

[17] The last survivors, who were very young at the time of the disaster, have passed away in recent years.

[18] "The earthquake shook down in San Francisco hundreds of thousands of dollars' worth of walls and chimneys. But the conflagration that followed burned up hundreds of millions of dollars' worth of property. There is no estimating within hundreds of millions

> There is no doubt that the fire was far more destructive than the earthquake, but it is also true that for reconstruction of the city, citizen and their leaders had to be sure to emphasize the gravity of the fire damage. [...] Still engineers and architects could not ignore what they had seen. Many remained committed to building structures that could resist both earthquake and fire. They tried to implement changes in building practices. Unnamed citizens, like my grandfather, remained concerned as well. (Tobriner, 2006: 197)

As a new actant (or a set of multiple actants) disasters disrupt and transform the pre-existing order of a particular space. If the potentiality of a disaster tends to be difficult to grasp, it is partly because acquaintance with large magnitude events has been muted for a long time. As pointed out in 1913 in the *Bulletin of Sociology of America*, "Shortly after the earthquake of April 1906, there was a general disposition that almost amounted to concerted action for the purpose of suppressing all mention of that catastrophe" (as quoted in Hansen, 1989: 117). In this period of recovery, catastrophic were outdated: San Francisco was doing well and made sure not to scare potential investors away.

In the next section, we explore the possibilities for integrating this multi-layered framework a step further in order to seek a theoretical framework that supports the co-construction of scientific and experimental knowledge. To do so, I focus our attention on the holds that connect earthquake experts to their research object and can be mobilized to describe the situation of "waiting for the Big One." Drawing both on James's pragmatist approach Actor-Network Theory and geography, we will define the nature and intensity of relations that exist between human and non-human actants that come together in situations of risk and disasters.

the actual damage wrought. Not in history has a modern imperial city been so completely destroyed. San Francisco is gone. Nothing remains of it but memories and a fringe of dwelling-houses on its outskirts. Its industrial section is wiped out. Its business section is wiped out. Its social and residential section is wiped out. The factories and warehouses, the great stores and newspaper buildings, the hotels and the palaces of the nabobs, are all gone. Remains only the fringe of dwelling houses on the outskirts of what was once San Francisco" (Tobriner, 2006: 197).

3.2 Hybrid Science, Hybrid Scientists

The earthquake risk is moving, because it is moved—moved by the motion of tectonic plates but also by mediated events that affect human and non-human actants. As Coen recalls, seismology has grown as a hybrid science, anchored in empiricism and relying on non-scientists to describe the phenomenon it needed to study.

3.2.1 Movement and Science

Historians working on disasters have contributed to nuance the understanding of what was once considered strictly "natural" or "factual" in many scientific fields, including in disaster studies (Bird, 1987; Demeritt, 2002; Favier & Granet-Abisset, 2009; Schenk, 2007). By documenting the transformations, disasters have operated in our categories of thinking since the European Middles Ages (Amador, 2004; Coen, 2013; Dynes, 1971, 1997).

> Making a science of disaster means constructing a basis for comparison, both geographic (how hard your town was hit versus mine) and historical (how much worse this one was than the one in your grandmother's stories). To this end, a science of disaster must constantly move back and forth between the natural and the social, the objective and the subjective, the global and the local. It must correlate geological formation and the built environment, instrumental data and human responses, planetary waves and local damage. Each informs the analysis of the other. (Coen, 2013: 6)

Coen's argument resonates with James' *Principles of Psychology*, first published in 1890, which focused on the definition of personhood,[19] and the way experiences shape the basis of any form of knowledge.

[19] James' perspective is very different from the one used by Freud in the years following James' *Principles of Psychology* (1890), followed by *Psychology: The Briefer Course*, in 1892. The relation between Freud and James is investigated well in a paper by Barbalet (2007). Also, though he recognized that Freud's work represents the future of research in psychology, he expressed reticence about what he calls Freud's "fixed ideas." James wrote, "I can make nothing in my own case of his dream theories and obviously symbolism is a most dangerous method" (as quoted in Barbalet, 2007: 41).

For James, "idea and knowledge are not the passive outcome of past experiences but [...] are the active projection of our experience into the future" (Barbalet, 2007: 18).[20] Barbalet explained the importance of this proposition: "this is a singularly Jamesian proposition, central to his pragmatism, not shared by his fellow founding pragmatists and not understood or appreciated by most of his followers. It is what brings his psychology and his pragmatic method together" (Barbalet, 2007: 18). In addition, the ambiguity of the future—or its uncertainty—is a source of distress, one that is both practical ("What can I do?") and philosophical ("What should I do?"), and that transforms the definition of rationality. In such a context, rationality enables people to "proceed in their particular affairs. Thus, James characterizes rationality in term of the particular emotional configuration that enables the actor to engage unknowable futures. [...] In this way, James characterizes rationality as a property of mind or a quality of action explicable in terms of its emotional qualities" (Barbalet, 2007: 22).[21] In recent years, there has been a renewed interest in James' work mostly focusing on the social dimension of his pragmatism.

Yet, Barbalet states that: "such compartmentalization should not be accepted, though, at the expense of the appreciation of the continuity of James thought" (Barbalet, 2007: 170). For Barbalet, "pragmatism was designated as a method, through which the embroidery of metaphysical speculation is unraveled. The purpose of this method is achieved by inquiring about consequences of practical actions that could be deduced from such speculations" (Barbalet, 2007: 17).[22]

Anthropologists working with ANT perspectives have also insisted on the fact that the subjective part of their scientific projects remains central in the realization of the latter. As Mialet (2012b) notes, scientists and scientific institutions have been on the edge of becoming schizophrenic,

[20] As we shall see later, this idea of a "mode of action" being, at the same time, active and passive is one of the main characteristics of the ANT concept of "attachment."

[21] As I shall discuss later, James' theory is important in decision-making theory. In most circumstances, we are forced to decide without knowing the outcome of the situation. It is what James called the "forced option," a concept which, as we shall see, is useful in the context of facing a risk.

[22] Barbalet also noted that James' last book, *Some Problems of Philosophy* (1911), unfinished at the time of his death, is a treatise on metaphysics, in spite of the anti-metaphysics of his first propositions. Even if we are not pursuing the same objective of dismissing metaphysics, I believe that James' account of emotions and social life is still noteworthy.

as they have tried to compose an image of science in accordance with its principle: a discipline definitively shaped by reason, and detached from individual components. This clear operation of distinction has been the basis of the construction of the scientific persona, as well as scientific discourse (Datson & Galison, 2010; Latour, 1991). However, Mialet claims:

> Thanks to Popper I knew that scientists had dreams and phantasms; from Merton I learned that they could be immoral; and I understood from my reading of Kuhn that they, like everyone else, were subject to emotional crises. But as far as these authors were concerned, what was given to the scientists (a certain form of humanity) was immediately taken away for the sake of science. Thus they invented elaborate systems to contain the scientist's subjectivity: hence for Popper, the distinction between the context of discovery (the realm of imagination) and the context of justification (the realm of method); for Merton, the distinction between "normal" institutions and "scientific" institutions ruled by universal norms; and for Kuhn, the distinction between a conflicted philosophical pre-paradigmatic science and the calm and orderly settled scientific paradigm, etc. (Mialet, 2012b)

In her work on the practices of invention, Mialet (1994) has defined what she calls a "distributed-centered subject" a materialized, collectivized, distributed and connected knowing subject (Mialet, 2012a: 192). Building on two fields of researches, one in an oil company and the other one studying the physicist Stephen Hawking, she noted that in a situation of making science, "the more (an) actor was linked to an institution, his object of research, his co-worker and so on, the more potential he had to become inventive, and the more inventive he became the more he seemingly distinguished himself as an inventor, as a kind of genius who existed beyond social material and cultural constraint" (Mialet, 2012a: 193). She added, recalling the multiple operations performed by scientists she has observed:

> This work of association, the evolution of this singular link, which is established between the actor and his object of research, is not only caused by phenomenons' of association. This link has a singularity of its own. And when we deploy it, we discover the capacities of one individual. […] The more he distributes, the more he becomes singular, the more he can operate transformations and news associations. In short, we are discovering a distributed actor closest to the actor from the psychology, thanks to his

> capacity to transpose problematic, to immerge himself in the objects and to metamorphose himself, then the actor network. We are discovering an actor equipped with a psychological plausibility, which moves us away from the cubist figure of the actor network. (Mialet, 1994: 290–291)

Taking into consideration each condition of experience in its multi-dimensionality,[23] Mialet addresses here the critic formulated by ANT researchers. Indeed, as noted by Nigel Thrift: "actor-network theory had tended to neglect specifically human capacities of expression, powers of invention, of fabulation, which cannot be simply gainsaid, in favor of a kind of flattened cohabitation of all things" (Amin & Thrift, 2007: 111). Thrift also draws the contours of a new definition of personhood, one that is no longer totally predicated, but can be deployed using the multiplicity of its links and connections.

3.2.2 Science and Dependence

Different theoretical frameworks can be attributed to different epistemologies, or "situated" perspectives, ideas, concepts, and methodologies. The scientific process has been built on the notion that distance between the subject engaged in study and the observed object would give the most reliable knowledge (Shapin & Shaffer, 2011). However, STS scholars have shown that this separation was never well defined (Fressoz, 2007; Latour, 1993). But how can we define the nature of the relation between earthquake experts and their subject of research when the former refer to themselves as Earthquake Junkies?

[23] "Philosophy has only ever generated differences by taking being qua being as a starting point (the Copernican revolution never happened: philosophy is still geocentric). It should be possible to adopt another position by 'trying out the Other'. This inquiry into the different ways of altering certainly has something empirical about it; in any case, it should stick as closely as possible to what is given in experience (in the full sense of the second empiricism, not the limited version of the first). The number of modes is greater than two, so we will ignore the subject/object dualism and call an end to the bifurcation of nature, not through going beyond it (that would only be counting to three) but through erasing it in a thousand different ways. The modes are of equal dignity; they are the product of a specific history—I would add of an historical anthropology—which does not aim to define a general ontology" (Latour, 2011: 316).

> K. calls us "Earthquake Junkies"! That's what he calls us. It's the Earthquake Junkies. The people who are, let's say very concerned, conscious, obsessed, whatever you want to call it, you know! [J.8]

The earthquake experts in the Bay Area—the Earthquake Junkies—share this hybridization of identity. They are also "distributed-centered subjects" who, in addition of the objects of their laboratory, take into consideration the place they live in. To understand the relation that the "Earthquake Junkies" maintain with their environment, I start by exploring the concept of distance.

In human geography, the approaches of the concept of distance have evolved from considerations about metric Euclidean space as to the construction of space as a result of "the intensity, nature and length of the relation that connects things together" (Lévy & Lussault, 2000: 269). Since the mid-nineteenth century, the Kantian conception of space was being challenged by mathematicians and physicists.[24] In the 1990s, Jonathan Murdoch, a geographer who was among the first to look at the possible interconnections between geography and ANT, gave a definition of distance that takes into consideration the relation between actants and affords understanding of the complex dimensions of social interactions.

> While it is no doubt true that "distance" remains tyrannical in the hands of some geographers, for most this beast has been rendered rather tame and pliable. [...] That is, ANT seeks to analyze how social and material processes (subjects, objects and relations) become seamlessly entwined within complex sets of association. This leads on to an interest in "network topologies", with the ways that spaces emerge as socio-material relations are arranged into orders and hierarchies. (Murdoch, 1998: 358, 359)

As Bingham and Thrift noted, this major recomposition replaced the Euclidian understanding of space as *one* possible conception amongst many others: "space and time proliferate [...] 'space' and 'time' are less important than the always unique acts of 'timing' and 'spacing' by which place-events are 'folded' and 'pleated' into existence" (Bingham & Thrift, 2000: 289–290).

[24]For instance, Gauss has "employed relative conception of space to show that 'activities and objects...define special field influence'...Thus spatial properties cannot be distinguished from object 'in' space and space itself can only be understood as a system of relation" (D. Harvey, as quoted by J. Murdoch, 1998: 358).

For geographers, actants interact in plurimodal ways: they are connected to one another and, as much as they are neatly linked, they also need to "keep their distance" to exist. In addition, distance, which here is conceived as spatio-temporal, alters actants: some resist and remain intact; others are transformed, never to be the same again. In this new redistribution, actants become "spatial operators," as they are shaped by space; their interactions can, in turn, transform space, but not necessarily always symmetrically.

A careful reader of ANT, Lussault developed what he called the "technology of distance", which he defined as "the system of the manifestation of distance and the social reality of its affects" (Lussault, 2007: 51). Lussault's spatial grammar starts at our fingertips and expands to the terrestrial globe, opening up to the notion of scales. First, our body envelope, understood as the physical boundaries of self; the space of life, the neighborhood related to everyday life (de Certeau, 1988). This specific space is where geographers working on the "non-representational theory" placed their analysis:

> We now understand that the space and rhythms of the everyday, everydayness, and everyday life [...] are not just a filigree bolstering an underlying social machine but a series of pre-individual ethologies that incessantly rehearse a materialism in which matter turns into a sensed-sensing energy with multiple centers. (Amin & Thrift, 2007: 17)

The third scale is the regional space: it is the space of displacement and transport of commodities, ideas, ideologies. The last is the world; a social space at the scale of the planet. Following the first hybridization of spatial theory with ANT, Lussault has argued that space can be defined as the "visibility regimen" of societal substances, where boundaries[25] hold things together.

As much as it seems obvious that "space" for a seismologist is not the same "space" as for an anthropologist, the idea of space also forms a poetic encapsulation of the latent world, well-structured and

[25] As much as this chapter proposes to take into consideration articulations across the interactions between space and "modes of existence," we should keep in mind that this is also because boundaries exist between the self and the world, between I and the others—thus, that, specifically, "otherness" can exist. In this sense, it is important that the reticular organization of the network also bears the mark of this reality.

pre-reflective. This interlacing of the pre-cognitive and cognitive dimensions of space is what provides the "paradox of space" that Thrift employs in his thinking:

> we all know that space is something lived in and through in the most mundane way – from the bordering provided by the womb, through the location of a coffee cup on our desk that is just out of reach, through the memories of building and landscape which intertwine with our bodies and provide a kind of poetic of space, through the ways in which vast political and commercial empires – and the resultant wealth and misery – can be fashioned from the mundane comings and goings of ships and trains and now planes, through to the invisible messages that inhabit the radio spectrum in their billion and etch another dimension of life. (Amin & Thrift, 2007: 17)

With the very artificial demand of researcher when doing field work. Thrift is calling for attention to a "social awareness", which involves a specific awareness for the different registers of human expression and their spacialisation.

This leads precisely to a necessity to pay attention to, and to be aware of, the actions that we perform in every instant of our everyday lives that we imported into the conceptual framework of this present work.

3.2.3 *Distance and Attention*

Defining what it is to wait for the Big One requires the expert to add another dimension to the Latourian network—namely, attention. This dimension is composed of the way's individuals living in a social environment organize, prioritize, store, and use information, which helps in understanding how emotions and knowledge interact using the concept of distanciation, "Distanciation in general refers to the stepping back or distancing of the observer or reader from an object of scrutiny. The process of distanciation. It has both a spatial and an emotional side" (Harvey, 1997). Of course, examining how people wait is an act of paying attention. Attention is an operation of the intellect that concerns the knowing subject. As William James defined it:

> Everyone knows what attention is. It is the taking possession by the mind, in clear and vivid form, of one out of what seem several simultaneously

> possible objects or trains of thought. Focalizations, concentration, of consciousness are of its essence. It implies withdrawal from some things in order to deal effectively with others, and is a condition, which has a real opposite in the confused, dazed, scatterbrained state that in French is called distraction, and *Zerstreutheit* in German. (James, 1890: 403–404)

Attention, in James's definition, implies a dimension of anticipation and, as such, it is opposed to "absent mindedness" (Depraz, 2013: 113). It is inscribed in a dynamic learning process and—just like waiting—is an active posture that requires that our senses, our minds, and our psychological wills be focused in the same direction and on a certain objective.

Attention is an operation of the intellect that concerns the knowing subject, but which takes root in the affects, experience, feelings and emotions. Attention is selective, and is defined as "reactive spontaneity":

> My experience is what I agree to attend to. Only those items which I notice shape my mind – without selective interest, experience is an utter chaos. Interest alone gives accent and emphasis, light and shade, background and foreground – intelligible perspective, in a word. (James, 1890 quoted by Depraz, 2013: 113)

James' understanding of "attention" implies the capacity to welcome novelty, and, in the case of earthquake risk, new forms of knowledge, agents, and subjectivities. Attention, as define by Isabelle Stengers, is an cognitive and intellectual operation that concerns the knowing and moral subject: "attention, which, for some, is a matter a free trade with the categories of obligations, which cannot be given away, without, in the same movement, being separated from what make them human" (Stengers, 2002: 28).[26]

For the researcher, but also for the expert and the resident, attention is the moment of care, the precaution at the moment of enunciation, which makes the difference between what Stengers calls "live knowledge" and "dead knowledge" (Stengers, 1999).[27] Attention—as its sister

[26] "Attention, ce qui, pour certains, relève d'un commerce libre par rapport aux catégories du devoir, de ce qui ne peut être enfreint sans que, par la même, ils se trouvent séparés de ce qui fait d'eux des humains" (Stengers, 2002: 28).

[27] In other writings, Stengers has called for our attention as a matter of civic duty: the necessary attention or awareness of our environment, which is a call echoed by authors concerning the question of political ecology and the consequences of the 2008 global financial crisis (see Hache, 2011; Serres, 2009).

concepts, perception and awareness[28]—has blurred conceptual boundaries. In the field where philosophers and psychologists have largely been involved in conceptual definitions, divergences appear, mostly on the degree of consciousness included in the notion. I consider that attention refers to active form: to be aware of something is to be able to acknowledge its existence. Attention brings together emotions and knowledge with what has been so important for the pragmatist—experience—and of course, the traces of past experiences.

3.2.4 *Distance and Attachment*

Hennion (2004a) noted the role of attention in the process of attachment, seeing it as a moment of suspension between two otherwise

[28] "James indicated that there are a great many categories of fringe experiences, not just one. However, he did not attempt an exhaustive list, or a systematic analysis of their relations to each other, or to other mental phenomena. He offered a few examples: 1. feelings of familiarity (p. 252). 2. feelings of knowing; e.g., as in the 'tip-of-the-tongue' experience (p. 251) 3. feelings of relation; these are subjective qualities associated with the relations between objects or ideas, as indicated by words such as 'and,' 'or,' 'if,' and 'but.' James says, 'We ought to say a feeling of "and"…quite as readily as we say a feeling of cold…' (p. 245). 4. feelings of action tendency; e.g., the intention to say so-and-so, just before it is articulated (p. 253). 5. attitudes of expectancy; the commands, 'wait,' 'look,' 'hark,' elicit distinct feelings of the domain from which a new impression is to come (p. 250). 6. feelings of 'rightness' or being 'on-the-right-track': this is a feeling that the content currently in the nucleus of awareness is congruent in some global way with our current goal structure (what James calls the 'topic' of our thought)" (Galin, 1994). The tip-of-the-tongue experience is particularly interesting, as it shows the progressive building up of the psychoanalytical theory between William James and Sigmund Freud. For James, "This state of our consciousness is peculiar. There is a gap therein: but no mere gap. It is a gap this is intensively active. A sort of wraith of the name is in it, beckoning us in a given direction, making us at moments tingle with the sense of our closeness, and then letting us sink back without the longed-for term. If wrong names are proposed to us, this singularity gap acts immediately to negate them…and the gap of one word does not feel like the gap of another, all empty of content as both might seem necessarily to be when described as gaps" (James, [1890] 1950: 251–252, quoted by Galin, 1994: 379). Galin notes that James' dual construction of awareness had limited impact between contemporary psychologists and cognitivists. The arguments against it have been the extremely limited time period during which the complex relationship between the fringes and the nucleus operate. A second argument has been the too-limited elements of definition between the fringe and the nucleus. Here, Galin argues that, "There must be some qualities more fundamental that definiteness that make the fringe information unsuitable for the purpose served by the nucleus information, and vice versa" (Galin, 1994: 13).

mundane, specific actions. As he began to investigate a theory regarding the network of attachment, Latour also developed what he calls "the middle voice."[29] Recusing a conception of action that would be defined only by either its passive or its active mode, bringing the question of emancipation or alienation to the fore, Latour states the importance of attachment:

> It is no longer a question of opposing attachment and detachment, but instead of good and poor attachment, then there is only one way of deciding the quality of these ties: to inquire of what they consist, what they do how one is affected by them. (Latour & Girard Stark, 1999: 22)

Going back to the scientists and experts of the Bay Area waiting for the Big One, we shall consider what James has defined as "The Sentiment of Rationality" where decision-making relies on the "force option." To illustrate this proposition, James gives the example of a rock climber who, when confronted with a difficulty, must trust herself enough to be able to succeed in a decisive leap. In this case, the positive emotion of trust could be as decisive as a negative emotion, which could lead to a fatal conclusion. Barbalet describes James' analysis in detail:

> In "The Sentiment of Rationality", James develops the point, that emotion constructs circumstances, through the case of the "Alpine climber" in which an actor's particular emotional commitment leads to a singular material outcome (James 1897, pp. 96–7). To escape serious difficulty, the Alpine climber must execute a dangerous leap that she has not performed before. If she is engaged by the emotions of confidence and hope, she is likely to perform a feat that would be otherwise impossible. Fear and mistrust, on the other hand, are likely to lead to hesitation, and this will increase the probability that the climber will miss her foothold and fall to her death. Whichever emotion is engaged will be commensurate with a particular outcome, but with contrastingly different consequences. The role of emotion in practical conduct or human agency, then, is to permit action that would be inhibited if it were to rely on logic or calculation alone. The evidence on which "rational" or deliberative calculation relies is simply not available for most social actions. The emotional contribution to agency is to overcome the uncertainty of the future by providing an emotional orientation to one possible future in the realization of a present

[29] The expression is borrowed from Emile Benveniste (1973).

> action. Otherwise action simply could not occur, and the actor would not be able to proceed. (Barbalet, 2007)

In the James example, context defines what the climber "ought" to do, and we see that emotion is the drive of that action. In a more contemporary version of this example, ANT researcher Antoine Hennion[30] applies a nuance to James's proposition, intending to demonstrate that, for the climber, the music amateur, the drug user, or any given individual, "we are actually concerned by a much larger range of practices and activities" (Hennion, 2007: 99). Moving toward the idea of the co-construction of action and the actor, Hennion likewise uses an example of a climber:

> The objective of the ascent conflates with the actual fact of climbing. What counts is entirely in what is happening. From the point of view of the subject, an analogous reduction erases the distinction between the goal and its realization. At the bottom of the route, the climber is eager to abandon all of the personal attributes that make up his regular identity. To practice this sort of thing together, one might begin at the outset by depositing all of that which is not concerned with this activity. [...] Before their cliff, the only important thing for a moment, for humans together like this, is common practice. As with all amateurs – be they the miniature modelers that come to compare their balsa wood planes on Sunday at the Bagatelle, or the Bocce players on the promenade in the South of France – it is the activity that defines them, not the inverse. (Hennion, 2007: 99)

Going one step further, to better define the connection between the risk expert and his or her object of research, November recalled that the climber metaphor has also been used by Bessy and Chateauraynault in their definition of expertise who discussed the intersection between the climbing hold, of "having hold over" the environment, but also to "give a hold" (Bessy & Chateauraynaud, 1995: 239).

3.2.5 *The Expert as Amateur*

Drawing on the figure of the "amateur" defined by researchers in ANT, experts and scientists in the Bay Area have recognized that their

[30] Hennion's work has long investigated classical music amateurs, while Gomart's studies have largely been concerned with drug users and the controversies about methadone programs (Gomart & Hennion, 1999; A. Hennion, 2004a, 2004b).

knowledge and their experience are interwoven, giving them certain responsibilities toward their fellow residents. “Amateurs learn to be affected,” recalled Annemarie Mol describing this important figure of the actor in ANT theory (Mol, 2010). But, “amateurs care about technicalities” (Mol, 2010), also. Building on Latour, Hennion has developed a sociology of attachments (Gomart & Hennion, 1999b); this is a sociology “in direct contact with things, uncertainty of sensations, method and techniques used to become sensitive to, and to feel the feeling of, the object being sought” (Hennion, 2007: 98).[31] The “amateur,” in Hennion’s vocabulary, is the co-constructed human actant of this relationship.

In the Bay Area, earthquake experts living on the fault lines are like climbers on the rock; they have to navigate between holds. Describing what they have called “the art of the hold,” Bessy and Chateauraynaud have shown that the experts are capable of identifying and deciphering the different modes of existence of the actants, and can, from their body space, activate the resources that are needed (Bessy & Chateauraynaud, 1995: 243).

Amateurs rely as much on the properties of objects—which, far from being given, have to be deployed in order to be perceived—as on the abilities and sensibilities one needs to train to perceive them; they rely as much on the individual and collective determinisms of attachment as they do on the techniques and devices necessary in a situation for things to be felt. Understood as reflexive work performed on one’s own attachments, the amateur’s taste, knowledge is no longer considered an arbitrary election that has to be explained by hidden causes. Rather, it is a collective technique, the analysis of which helps us understand the way we make ourselves available, to things, and to moments, while simultaneously controlling how those feelings may be shared and discussed with others (Hennion, 2007: 98).

What interested Mialet, Chateauraynaud and Bessy, Gomart, Hennion, November, and Latour is the study of the devices, practices, and interests that work together to create the inventor, the drug user, the amateur musician, or the expert; it is this type of relationship that

[31] Hennion firmly inscribes this in his opposition to a sociology of culture, for which these moments “are directly denounced as ritual whose principal function is less to make amateurs ‘feel’ than to make them ‘believe’” (Hennion, 2007: 98).

explains the development of attachment.[32] Following the form of network, the researcher can trace concomitant attachments: "the formidable proliferations of objects, properties, beings, fears, techniques that make us do things unto others" (Latour, 2009: 24).

The introduction of these non-rational dimensions of our lives, our processes of decision-making and our production of knowledge, ultimately opens up new perspectives in the definition and organization of our individual worlds. It allows existences—or ontologies—of actants to come into being. Our attachments become visible, the material strings of the network—things, thoughts, beliefs, and experiences that matter—allow us to put each into perspective (Bennett, 2004; Hache, 2011).

3.3 A Network of Attention to the Risk

Risk is never a clear given; it needs to be comprehended in its full complexity, to be approached with caution, considering the many ways in which residents, concerned citizens, and experts are actively waiting for the Big One. If a consensus has emerged over the years about the seismicity of the region, controversies remain about how best to frame and define past events, and how to manage their legacies. Deploying this first layer of complexity around an understanding of what precisely the earthquake risk is for the area, we can then identify how controversies have emerged, and how historical research can illuminate certain aspects of this subject.

This chapter has focused on some of the possible "readings" of the concepts mobilized in the following chapters of this work: space, risk, and attention. We have argued that the articulation of these three concepts allows us to look at the ways in which the likelihood of earthquake is defined as a risk and, as such, how it is memorized and forgotten, perceived and experienced, understood and studied in the Bay Area.

While waiting for the Big One, experts and residents are looking at the mediations and actants that create a network, and, as we have seen,

[32] First developed by Latour in 1999, the quotation of attachment has been comically illustrated by the Spanish artist Quino. In his drawing, an emblematic little girl, Mafalda, says to her father, who is smoking a cigarette, "I thought the cigarette was smoking you." The humor here is based on the question of relations, according to the dialectical construction of being either active or passive; meaning that one is either the master of an object or is dependent on it.

this is a space that will, perhaps, allow residents of the San Francisco Bay Area to be resilient in the event of a catastrophic event. Risk and disaster stand at the intersections between territory, science, and experience, and between human and non-human actants, or what Pratt (1992) has called a "contact zone." The space of risk is a zone of contact, where human and non-human elements interact, and where different dimensions of experience are combined. Interstices between the conditional and the future, between the definition of a risk and the envisioning of a catastrophe, between a regimen of action and a regimen of attachment, are also what characterize the relationships that earthquake experts maintain with the very real risk of an earthquake occurring in the Bay Area.

Not a risk takers, not potential victims, and not experts detached from grounded experience, earthquake experts in the Bay Area have developed a particular connection to the object of their research and refer to themselves as Earthquake Junkies. As actants "in between," as in the famous image of the climber on the rock, earthquake experts developed a certain form of attachment toward the place where they live. This singularity of scientists and experts is what make the situation of Bay Area earthquake research so unique. In this "contact zone"—between tectonic plates, science, experiences, and so forth—earthquake experts have learned how to deal with the multiple actants. They have learned to "think as" the earthquake—just as Mialet's engineers have needed to think "as the oil" (Mialet, 1994)—but they have also learned how to think as individual residents—those who live their lives above an active seismic fault. They have not only learned about the complex mechanisms that can trigger an earthquake, but also the equally numerous complex policies that would make the Bay Area more resilient.

As we shall see, the experience of living with earthquake risk deeply transforms our understanding of the world. When they move, seismic faults displace lines: not only the lines of experience of dwelling, but also more epistemological and metaphysical lines—the lines of thoughts that are often used to define the word.

To understand the instauration of earthquake risk in the Bay Area, we shall look at two corpuses of literature that have, too often, been considered separately: that dedicated to risk studies and that to disaster studies. In the last few decades, frustrations have been expressed among the community of risk and/or disaster researchers about this segmentation, which has made the full understanding of this complex field a significant challenge (November et al., 2009; Cutter, 2005: 40). Pigeon noted:

"Not only does segmentation seems simplistic and odd in light of field experiences, but it tends to be validated by the dual hazard-vulnerability which favors it strongly, in spite of increasing [...] questions of coherence" (Pigeon, 2005: 26). To appreciate this complaint fully, we bring together these lines of researches.

Working in the Cold War era, American social scientists were urged by a US government that feared a nuclear attack on American soil to study the consequences of such events on populations. These researches were the first attempt to address the complexity of disaster in a systematic way, focusing on the distributions of facts, values, and both human and non-human actants as aggregated by pioneers in this field. We shall see how other approaches of risk and disaster have also tackled the problem, tracing an epistemological "continuum" that is useful in understanding the different perspectives that have been taken to frame and define earthquake risk. Finally, we shall see how, following James, a pragmatic approach to earthquakes could offer the best framework with which to consider the complexity of living with the risk of a disaster.

Following James and Bruno Latour, we will deploy the definition of the earthquake as a "living agent", or "the collective *name* of all the cracks and shakings and disturbances that happen" (James, 1906). I will also discuss how integrating the new actants changes the definition of science and the understanding of this scientist, experts and define a hybrid scientist—an expert who is emotionally and rationally connected to his object of research.

References

Adler, P. (1992). *Fire in the hills*. Auto-published.

Amador, F. (2004). The causes of 1755 Lisbon earthquake on Kant. *Actas VIII Congreso de La Sociedad Espanola de Historia de Las Ciencias Y de Los Tecnicos*, 485–495. Retrieved from http://dialnet.unirioja.es/servlet/dcfichero_articulo?codigo=1090088.

Amin, A., & Thrift, N. (2007). Cultural-economy and cities. *Progress in Human Geography, 31*(2), 143–161. https://doi.org/10.1177/0309132507075361.

Barbalet, J. (2007). Classical pragmatism, classical sociology: William James, religion and emotion. In P. Baert & B. Turner (Eds.), *Pragmatism and European social theory* (pp. 20–21). Oxford: The Bardwell Press.

Bennett, J. (2004). The force of things: Steps toward an ecology of matter. *Political Theory, 32*(3), 347–372. https://doi.org/10.1177/0090591703260853.

Benveniste, E. (1973). Problems in general linguistics. In *Problems in general linguistics* (p. 317). Miami: University of Miami Press.

Bessy, C., & Chateauraynaud, F. (1995). *Experts et Faussaires*. Paris: Métailié.

Bingham, N., & Thrift, N. (2000). Thinking space. In M. Crang & N. Thrift (Eds.), *Thinking space* (pp. 281–301). London: Routledge.

Bird, E. A. R. (1987). The social construction of nature: Theoretical approaches to the history of environmental problems. *Environmental Review, 11*(4), 255–264.

Coen, D. (2013). *The earthquake observers: Disaster science from Lisbon to Richter*. Chicago: The University of Chicago Press.

Cutter, S. (2005). Are we asking the right questions? In R. Perry & E. L. Quarantelli (Eds.), *What is a disaster? New answer to old questions* (pp. 39–49). Philadelphia: Xlibris.

Datson, L., & Galison, P. (2010). *Objectivity*. Cambridge: Zone Books.

Davis, M. (1998). *Ecology of fear: Los Angeles and the imagination of disasters*. New York: Vintage Books.

de Certeau, M. (1988). *The practice of everyday life*. Berkeley: University of California Press.

Demeritt, D. (2002). What is the "social construction of nature"? A typology and sympathetic critique. *Progress in Human Geography, 26*(6), 767–790. https://doi.org/10.1191/0309132502ph402oa.

Depraz, N. (2013). Attention et Affection: la micro-genèse husserlienne de l'attention à la lumière des perspectives empiriques de Stumpf, James, de Külpe et de Titchener. In *Phenomenology 2005, Vol. 3: Selected essays from Euro-Mediterranean area, Part 1* (English, Spanish and French Edition, pp. 103–135). Bucharest: Zeta Books.

Dynes, R. R. (1971). *The dialogue between Voltaire and Rousseau on the Lisbon earthquake: The emergence of social science view* (No. DE 19716).

Dynes, R. R. (1997). *The Lisbon earthquake in 1755: Contested meaning in the first modern disaster* (No. #255).

Favier, R., & Granet-Abisset, A.-M. (2009). Society and natural risk in France, 1500–2000: Changing historical perspectives. In C. Mauch & C. Pfister (Eds.), *Natural disasters cultural response* (pp. 103–136). Lanham: Lexington Book.

Fressoz, J. (2007). Beck back in the 19th century: Towards a genealogy of risk society. *History and Technology, 23*(4), 333–350. https://doi.org/10.1080/07341510701527419.

Galin, D. (1994). The structure of awareness: Contemporary applications of William James' forgotten concept of fringe. *Journal of Mind and Behavior, 15*(4), 375–400.

Genthes, A. (1936). *As I remember*. Retrieved from http://www.zpub.com/sf/history/ag/ag-asi-10.html.

Gilbert, C. (2007). Crisis analysis: Between normalization and avoidance. *Journal of Risk Research, 10*(7), 925–940. https://doi.org/10.1080/13669870701504731.
Gomart, E., & Hennion, A. (1999a). A sociology of attachment: Music amateurs, drug users. *The Editorial Board of The Sociological Review*. https://doi.org/10.1111/j.1467-954X.1999.tb03490.x.
Gomart, E., & Hennion, A. (1999b). A sociology of attachment: Music amateurs, drug users. In J. Law & J. Hassard (Eds.), *Actor network theory and after* (pp. 1–28). Oxford: Wiley-Blackwell.
Hache, E. (2011). *Ce à quoi nous tenons. Propositions pour une écologie pragmatique* (Les empech). Paris: La Decouverte.
Hansen, G. (1989). *Denial of disaster: The untold story and photographs of the San Francisco earthquake of 1906*. San Francisco: Cameron & Company.
Harvey, D. (1997). *Justice, nature and the geography of difference*. Oxford: Blackwell.
Heller, A. (2006). *Re-creating the 1906 San Francisco earthquake*. Retrieved from https://www.llnl.gov/str/Sep06/Rodgers.html.
Hennion, A. (2004a). Pragmatics of taste. In M. Jacobs & N. Hanrahan (Eds.), *The Blackwell companion to the sociology of culture* (pp. 131–144). Oxford: Blackwell.
Hennion, A. (2004b). Une sociologie des attachements. D'une sociologie de la culture à une pragmatique de l'amateur. *Sociétés. Pratiques Musicales, 3*(85), 9–24.
Hennion, A. (2007). Those things that hold us together: Taste and sociology. *Cultural Sociology, 1*(1), 97–114. https://doi.org/10.1177/1749975507073923.
Hoffman, S. (1998). Eve and Adam among the embers: Gender patterns after the Oakland Berkeley firestorm. In E. Enarson & B. Hern Morrow (Eds.), *The gendered terrain of disaster: Through women's eyes* (pp. 55–61). Westport, CT: Greenwood Publishing.
James, W. (1890). *Classics in the history of psychology. Henry Holt and Company*. Retrieved from https://psychclassics.yorku.ca/James/Principles/prin11.htm.
James, W. (1906). On some mental effects of the earthquake. *The Youth's Companion*. Reprinted in James, H., Jr, 1911, *Memories and Studies* (H. James, Jr.) Longmans, Green, & Co. (Vol. June). New York.
Keeley, J. (2005). Fire history of the San Francisco East Bay region and implications for landscape patterns. *International Journal of Wildland Fire, 14*, 285–296.
Kingston, M. H. (2003). *The fifth book of peace*. New York: First Vintage International Edition.
Latour, B. (1991). *Nous n'avons jamais été modernes. Essai d'anthropologie symétrique* (Poche). Paris: La Découverte.

Latour, B. (1993). *The pasteurisation of France*. Cambridge: Harvard University Press.

Latour, B. (2005). What is the style if matters of concerns. In *Spinoza Lectures* (p. 27).

Latour, B. (2009). *Sur le Culte Moderne des Dieux Faitiches*. Paris: Les Empecheurs de Tourner en Rond, La Decouverte.

Latour, B. (2011). Reflections on Etienne Souriau's Les differents modes d'existence. In G. Harman, L. Bryant, & N. Srnicek (Eds.), *The speculative turn continental materialism and realism* (Anamnesis, pp. 304–334). Melbourne: re.press. Retrieved from http://books.google.com/books?id=coBDqJQeAQYC&pg=PA304&lpg=PA304&dq=Latour+Souriau&source=bl&ots=wHqRXjOttu&sig=GjPBBQjrUsrgrRgvWeykV9m6Q7Q&hl=en&sa=X&ei=XDFnUe_ZNoXmiwK7toDADQ&ved=0CGIQ6AEwAw#v=onepage&q=LatourSouriau&f=false.

Latour, B., & Girard Stark, M. (1999). Factures/fractures: From the concept of network to the concept of attachment. *RES: Anthropology and Aesthetics, 36*(Autumn), 20–31. Retrieved from http://www.bruno-latour.fr/sites/default/files/downloads/76-FAKTURA-GB.pdf.

Lévy, J., & Lussault, M. (2000). *Logique de l'espace, esprit des lieux. Géographie à Cerisy*. Paris: Belin.

Lussault, M. (2007). *L'homme spatial, la construction sociale de l'espace humain*. Paris: Editions du Seuil.

Mialet, H. (1994). *Le Sujet de l'invention. Etude empirique de la conception d'une idee neuve: comparaison des methodes philosophique et sociologique*. Sorbonne Paris I.

Mialet, H. (2012a). *Hawking incorporated: Stephen Hawking and the anthropology of the knowing subject*. Chicago and London: University of Chicago Press.

Mialet, H. (2012b). Where would STS be without Latour? What would be missing? *Social Studies of Science*. https://doi.org/10.1177/0306312712440311.

Mitchell, M. (2011). *Fire ruin renewal*. The Phenix Firestorm Project. Retrieved from http://www.margarettamitchell.com/firestorm/.

Mol, A. (2010). Actor-network theory: Sensitive terms and enduring tensions. *Kölner Zeitschrift Für Soziologie Und Sozialpsychologie, 50*(1), 253–269.

Murdoch, J. (1998). The spaces of actor-network theory. *Geoforum, 29*(4), 357–374. https://doi.org/10.1016/S0016-7185(98)00011-6.

November, V., Penelas, M., & Viot, P. (2009). When flood risk transforms a territory: The Lully effect. *Geography, 94*, 189–197.

Pigeon, P. (2005). *Géographie Critique des Risques* (Antropos). Paris: Economica.

Pratt, M. L. (1992). *Imperial eyes: Travel writing and transculturation*. London: Routledge.

Routley, G. J. (1991). *The East Bay Hills fire Oakland-Berkeley, California USFA-TR-060*. Oakland-Berkeley, CA. Retrieved from http://www.usfa.fema.gov/downloads/pdf/publications/tr-060.pdf.

Schenk, G. J. (2007). Historical disaster research, state of research, concepts, methods and case study. *Historical Social Research, 32*(3), 9–31.

Schiewe, J. (2011, October 19). *After Oakland Hills fire, residents build of-the-wall homes.* Oakland North. Retrieved from http://oaklandnorth.net/2011/10/19/after-oakland-hills-fire-residents-build-off-the-wall-homes/.

Self, R. (2005). *American Babylon: Race and the struggle for postwar Oakland.* Princeton: Princeton University Press.

Serres, M. (2009). *Temps des crises* (Manifestes). Le Pommier.

Shapin, S., & Shaffer, S. (2011). *Leviathan and the air-pump: Hobbes, Boyle, and the experimental life.* Princeton: Princeton University Press.

Solnit, R. (2009). *A paradise built in hell: The extraordinary communities that arise in disaster.* London: Viking.

Stengers, I. (1999). Savoirs Scolaires : qu'est-ce qui est digne d'être enseigné? In Conseil de l'Europe (Ed.), *La responsabilité, du principe aux pratiques Delphes.* Retrieved from https://books.google.ch/books?id=BP5dhvqJ9TIC&pg=PA95&lpg=PA95&dq=savoirs+vivant++isabelle+stengers&source=bl&ots=1qsblPBk2S&sig=ACfU3U2FjwSthHnhkBfCUTkq4DppJJ7w-6w&hl=en&sa=X&ved=2ahUKEwiclrSNhd_hAhVU8aYKHfwuDyQQ6AEwCHoECAkQAQ#v=onepage&q=savoirsvivantisabellestengers&f=false.

Stengers, I. (2002). Un engagement pour le possible. *Cosmopolitiques, 1,* 27–36.

Taussig, M. (2009). Culture of terror—Space of death: Roger Casement's Putumayo report and the explanation of torture. *Comparative Studies in Society and History, 26*(3), 467. https://doi.org/10.1017/S0010417500011105.

Taylor, T. (2011). 20 years after the fire, revelation and tribute. *The New York Time.* Retrieved from http://www.nytimes.com/2011/10/21/us/richard-misrach-reveals-his-images-of-oakland-berkeley-fire.html?_r=0.

Thrift, N. (2008). *Non-representational theory, space, politics, affect (International Library of Sociology)* (J. Urry, Ed.). London: Routledge—Taylor & Francis Group.

Tobriner, S. (2006). *Bracing for disaster. Earthquake-resistant architecture and engineering in San Francisco, 1838–1933.* Berkeley: The Bancroft Library, University of California Berkeley.

Walker, R. (1995). Landscape and city life: Four ecologies of residence in San Francisco Bay Area. *Cultural Geographies, 2*(1), 33–64. Retrieved from http://oldweb.geog.berkeley.edu/ProjectsResources/Publications/BayLandscape.htm.

Watts, M. J., & Bohle, H. G. (1993). The space of vulnerability: The causal structure of hunger and famine. *Progress in Human Geography, 17*(43), 43–67. https://doi.org/10.1177/030913259301700103.

Wisner, B., Blaikie, P., Cannon, T., & Davis, I. (2006). *At risk.* London: Routledge.

CHAPTER 4

Living with Risks

Myths about earthquake talk about the metaphysical at a particular time. They let us know about the organization of the world, and how facts and values are constructed and cared for.[1] In the Bay Area of San Francisco, little is known about the Ohlone[2] tribes (Margolin, 2003) the first inhabitants of this region of California, and their relationship to earthquakes. The following story has been carry through generation and is now a well known tale of creation.

> At that time, the entire world was covered with water, except for two islands, one of which was Monte Diablo and the other of which was Reed Peak. There was a Coyote on Reed Peak. [...] One day Coyote saw a feather floating on the water, and, as it reached the island, is suddenly turned into an Eagle. Spreading its broad wings, the Eagle flew up onto the mountain. Coyote was much pleased with his new companion, and they lived together in great harmony. [...] Then they consulted with each other, and decided to make human beings. [...] Soon the first human beings had children, and the level of the water went down so that there was

[1] There are many tales and myths about earthquakes which are mainly used today as sources of examples of false belief. As we shall see later, the frontier between knowledge and beliefs has been evolving following the solidification of seismology. Ancient myths often involve deified animals such as the frog in China, catfish in Japan, elephant in India, or turtles in Native and Latin American traditions.

[2] The group of tribes called the Ohlone was one of the densest Indian populations north of Mexico, most of them living in the coastal area between Point Sur and San Francisco.

C. Mazel-Cabasse, *Waiting for the Big One*,
https://doi.org/10.1007/978-3-030-15289-5_4

> more land for the human beings […] and the more human beings there were, the more the waters decreased, until at least where there was dry land where there once had been water. At that time, what is now known as the Golden Gate was a chain of mountains, and you could walk from one side to the other side without getting your feet wet. The water that came down from the east had to go out through some other rivers somewhere. But then a great earthquake struck, and chain of mountains was cut in two, forming what we now call the Golden Gate. Then the waters of the Great Ocean and the Bay could at last come together, and the land became as we now know it.[3] (Harper, 2010)

The myth tells us about the closed interaction between men and their environment. Instead of census data, unreinforced masonry building, industry exposure databases, and estimations of the interruption costs to business, it brings together a very different form of collective. The spatial and temporal dimensions, mobilize mountains, sea, sky and animal beings, the size of a foot, to create a very different sort of collective.

The everydayness of the risk of disaster, the close attention to the details and the relations between the earthquake phenomenon and apparently mundane objects can be grasped in a different way. Scientific scenarios have focused on the elaboration of the concept of risk, defining not only the chains of actants that will create the earthquake, but also its consequences while fictional writers have borne witness to the more individual, more personal dimensions of the disasters.

Imaginary destruction of the city structures a narrative around the disappearance of one world. Fictional stories about the Big One remind us of its transformative dimensions: risk the brutal destruction in slow motion. Disaster movies and novels (Taylor, 2006) picturing this destruction may be good starting points from which to question how earthquakes change our relation to the world. In his novel *The Big*

[3]Note from the author: "The above tale is adapted from 'Tradition of the California Indians,' by H. B. D., in *Hesperian Magazine*, vols. 2–3 (ed. F. H. Day, San Francisco, vol. III, no. 1, September, 1859), p. 326. H. D. B. says this tale came 'from the lips of one of our most venerable pioneers, and I give it as I heard it.' This tale is cited by Hubert Howe Bancroft in his *Native Races*, vol. 3 (San Francisco: History Company, 1886), p. 88; and by Alfred Louis Kroeber in his *"Myths of South Central California" American Archaeology and Ethnology Shoshonean Dialects of California*, vol. 4, no. 3 (Berkeley: University of California, 1907), pp. 188–189. I have made a few changes based on Kroeber. It is impossible to say now exactly which Ohlone people this story came from, but it would be a people who knew Monte Diablo and Reed Peak, whichever mountain Reed Peak once referred to."

One, Littlejohn looks at several protagonists before, during and after the earthquake. For each protagonists, the earthquake is a turning point, around which personnal stories are articulated. In the following excerpt, the first moments of Big One throw the character in a state of mental confusion, which will remain the turning point of his future relation the world.

> As Mike stared out the still uncurtained window, he heard it crack. Then the floor began rocking and leaping as if he were on a wooden boat in a storm. Crashing noises all around him inside the building preceded his awareness of the noises beyond thunder outside, which made it sound as if the city were being suddenly attacked by explosions from under earth. A terrorist attacks. No. The floor was still pinching, the walls were still shaking violently to and fro. This was it. The earthquake the paper and TV had been predicting all his life. The Big One. (LittleJohn, 2011)

4.1 Seizing the Earthquake as a Phenomenon

On October 20, 2011, at 2:41 p.m. Pacific Standard Time, just a few hours after the "Great California Shake Out," a voluntary-based earthquake drill, an M4.0 earthquake rattled the San Francisco Bay Area; followed by an M3.8 aftershock at 8:16 a.m. The coincidence of the two events, the earthquake drill and the actual earthquake, generated amused comments as well as real concerns about the seismic "status of the region." After all, the 2010 Haitian and 2011 Japanese earthquakes were still very present in everybody's mind. Traditional media outlets were buzzing and, in the social sphere, tweets were echoing street conversations. On 27 October, another earthquake hit, a more modest M3.6, directly underneath the University of California, Berkeley campus. Though the shock was low in intensity, it was still clearly felt. One seismologist noted, "These are gentle reminders that we live in earthquake country" (Sanders, 2011).

4.1.1 "Did You Feel It?"

The perception of the floor, walls, and the surroundings, all moving, the ground falling away under one's feet, along with a definite feeling of spatial disorientation that takes place: an earthquake is happening. The feeling of "solid ground" moving is deeply unsettling. While hands are gripping something, the idea of being confronted to an earthquake slowly makes its way through the nervous system. "Earthquake!"—but then, "How big?" A century earlier, in a post-earthquake account, the philosopher William James described his own experience of the famed 1906 San Francisco earthquake, and recalled a California friend's warning about the possibility of a seismic event:

> Accordingly, when, lying awake at about half past five on the morning of April 18 in my little "flat" on the campus of Stanford, I felt the bed begin to waggle, my first consciousness was one of gleeful recognition of the nature of the movement. "By Jove," I said to myself, "here's B's old earthquake, after all!" And then, as it went crescendo. "And a jolly good one it is, too!" I said.
>
> Sitting up involuntarily, and taking a kneeling position, I was thrown down on my face as it went fortior shaking the room exactly as a terrier shakes a rat. Then everything that was on anything else slid off to the floor, over went bureau and chiffonier with a crash, as the fortissimo was reached; plaster cracked, an awful roaring noise seemed to fill the outer air, and in an instant all was still again, save the soft babble of human voices from far and near that soon began to make itself heard, as the inhabitants in costumes negligés in various degrees sought the greater safety of the street and yielded to the passionate desire for sympathetic communication. (James, 1906)

Taking a broad view, earthquakes are what happen when familiar categories lose their everyday, common properties moved without warning. It is a moment where the well-established ordinance of the world as we know it, is transformed. Objects, time, values, space, thinking, and emotion: everything changes substance. Every "thing" becomes a mass, moved by gravity; and the human body is one of them. Of course, the process of a rumbling earthquake is usually very quick, often not lasting more than couple of seconds.[4] But these few seconds can be life-changing. Writing to his brother Henry after earthquake, James declared: "[It is] impossible not to feel it as animated by a will, so vicious was the expression of the temper displayed, and I see now how absolutely inevitable was the primitive theological interpretation of such disturbance"

[4] Very occasionally, however, they can be surprisingly long; the Tohoku earthquake, for example, lasted approximately six minutes.

(Livingston, 2012)—a disturbance so large that it also impacts the categories of human and non-human, physic and meta-physic.

Like counting the seconds between the sound of thunder and the sight of lightning to guess how far off the storm is, children in "earthquake country"[5] often learn to count between seismic waves and make guesses about the location of the epicenter. The experience of an earthquake depends not only on its distance from the epicenter, but also on the crustal material that seismic waves must travel through. For the observer, the feeling of an earthquake also depends on the type of building they are standing in, and the intensity—or the quality of—their attention to the phenomenon. Some earthquakes feel like a sudden, sharp jolt, which is then followed by strong shaking. In cases of extremely violent events, as evoked by James's recollection, the shaking can be so strong that it throws everything and everybody to the ground. A nearby earthquake will most often feel "sharp" and pass quickly, while distant ones usually generate more shaking that last longer.[6]

Direct experience of earthquakes is one of the ways of knowing what it is to live in a seismic zone.[7] Today,[8] coverage in local newspapers focuses on witness testimonies and their descriptions of the intensity, sensations, and consequences of the event within their communities. On October 20, 2011, on Telegraph Avenue in Berkeley, an employee at a store said, "[It was] a real quick one, like a bunch of jerking back and forth. It was one strong tremor" ("Berkeley Earthquake: Much Chatter About 3.9 Tremblor, but Little Damage," 2011). On the UC Berkeley campus, a student recalled, "It was like a sudden jolt, as if someone were to push you from behind. It was a pretty big shake"

[5] The expression comes from The Earthquake Country Alliance, "a public-private partnership of people, organizations, and regional alliances that work together to improve preparedness, mitigation and resiliency" in Southern California, the Bay Area, and the Red Wood Coast (http://www.earthquakecountry.org/).

[6] Comparing experiences can fuel conversations for a while. For many, it is often interesting to think about and discuss what they experienced: the growing rumble of the pressure waves or the shock of the secondary waves.

[7] Of course, the human body does not perceive all earthquakes, and the experience of an earthquake can be indirect. But very small earthquakes, such as those at M2.5 and lower, which may not be perceived by the human body, are recorded and are visible on the Real Time USGS maps. Therefore, tremors that are neither humanly perceived nor recorded through instruments do not "exist" as earthquakes.

[8] As we shall see later, earthquake accounts have a long journalistic tradition.

("Berkeley Earthquake Hits During 'Shakeout' Drill," 2011). In a local supermarket on Shattuck Avenue, an employee noticed that the earthquake was not big enough to throw merchandise down from the shelves. Another person, a new resident to the area who had been visiting friend when the earthquake happened, recalled "I was surprised to hear people screaming."[9]

The experience of an earthquake—how it felt, where it had occurred, and what damage it caused—comprises personal experiences, scientifically considered as facts. The "Did you feel it?" program (DYFI) by the USGS collects real-time information and measurements from common, non-scientific observers of an earthquake. "The idea of the DYFI program is that citizens use an Internet Web site (http://earthquake.usgs.gov/dyfi/) to report their experiences and observations for any earthquakes that they have felt (or not felt) by answering a simple multiple-choice questionnaire" (Atkinson & Wald, 2007).

Respondents' answers are used as a diagnostic of Modify Mercalli Intensity at the observer's location and later visualized into a map. With the help of the "distance versus intensity" calculation, these personal testimonies are translated into a Community Internet Intensity Map (CIIM). The CIIM records perceptions of earthquakes, organizes information, and helps scientists to visualize experiences derived from collective feelings, perceptions, and observations of the event. Also called "felt-maps," these community generated maps are very valuable for the scientific community, "especially when considering the limited efforts required for implementation" (Bossu et al., 2011). But, as the authors noted, the felt-map can only be used in:

> areas with significant Internet usage and where there are well-known Web sites providing earthquake information. Also, the ability to tie IP addresses to locations varies with the network infrastructure setup in different countries. The approach complements online macro seismic surveys [e.g., Wald et al., 1999]—where users are invited to fill out a questionnaire—by providing less accurate (the level of damage is not characterized) but more rapid (10–20 minutes compared with 1 hour to a few hours) information on the effects of the earthquake. (Bossu et al., 2011)

[9] C.36, personal communication, October 2011.

The Community Internet Intensity Map program first went online in 1997 to assess the scope of earthquake frequencies and impacts of particular seismic events, especially in areas not well-covered by the system of seismic stations (Fig. 4.1).[10] During the 2011 California earthquakes, the number of individual, personal responses often exceeded the number of seismic stations and considerably increased the potential of scientific clarity. For example, more than 15,000 persons reported their observations within three hours of the October 20, 2011 quake ("Deep Under Berkeley," 2011).

"Feeling" of an earthquake designs the set of empirical observations emotion the map scales the characteristics of the shaking measured on a level from "weak" to "extreme."[11] As a tool, the map allows the representation of an event and an easier way to record, archive, and display information: they monitor normal—one could say "healthy"—crustal activity in a seismic region.

With time, the USGS's interfaces have changed the way residents understand and react to earthquakes, as well as the way scientists consider a layman's account.[12] With the recent increase in portable and interconnected devices such as cell phones, computer tablets, and laptop computers, sharing information about earthquakes between individuals, the media, and professional seismologists has been transformed. Not long ago, people would turn to the radio or television after a tremor and wait to learn more; today, they connect to the USGS website to help create the CIIS map: it making clear that, as an object of science, an earthquake

[10]Modified Mercalli Intensity Maps, which are the precursors to CIIM, have been used in the USA since 1931. "Typically, this is done by collecting responses to a postal questionnaire that is sent to each post office near the earthquake, and to a sparser sample of post offices with increasing distance from the earthquake. This way of preparing a seismic intensity map can take months to complete" (Wald, 2001).

[11]Also, the Internet questionnaire collects certain information regarding observers' emotional states; thus far, this information has not been taken into consideration when building CIIS maps.

[12]In Chapter 5, I show how these maps have also transformed the relationship between scientists and non-scientists regarding earthquake information, generating more data with greater accuracy: "To date, more than 750,000 responses have been compiled in the United States alone. The DYFI data make up in quantity what they may lack in scientific quality and offer the potential to resolve longstanding issues in earthquake ground-motion science" (Atkinson & Wald, 2007, 362).

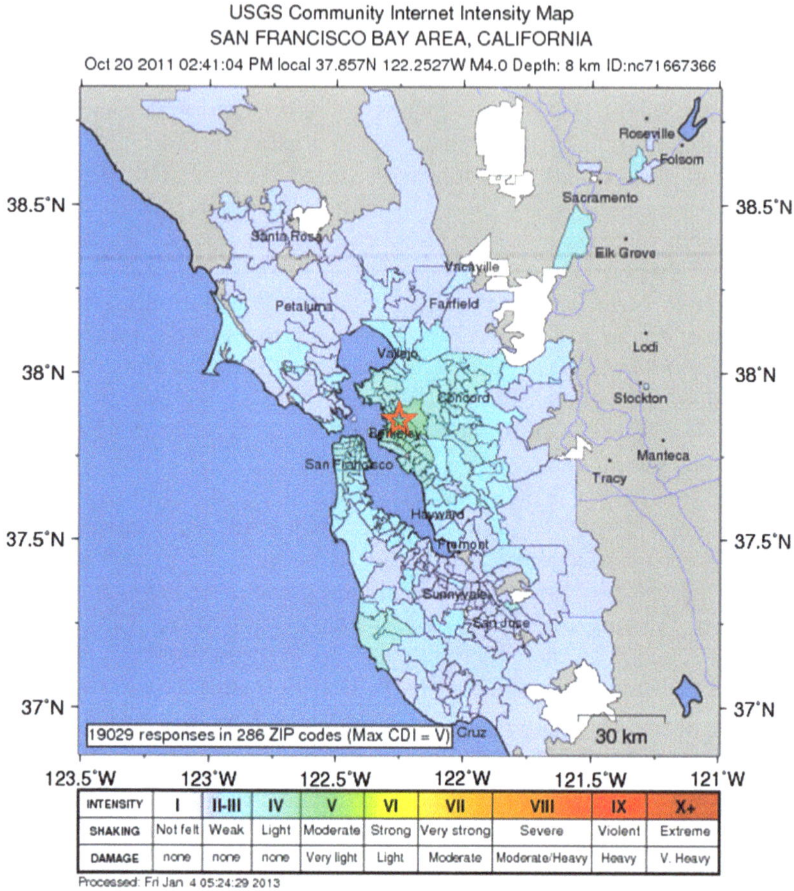

Fig. 4.1 USGS Community Internet Intensity Map *Source* USGS (2011)

needs to be experienced to be understood; it needs to be felt, to be known.[13]

[13]As I shall discuss later, if the most recent technologies used for early warning systems are now able to give an estimation of earthquake intensity using the analysis of energy emitted by its first waves, in 2019 the early warning system is not yet in place in California. It also needs to be noted that, during the Tōhoku earthquake, the Japanese early warning system did not provide adequate information regarding this earthquake's intensity.

4.1.2 *Seeing the Quake: The Indirect Experience from Elsewhere*

Seen historically, San Francisco is a city of travelers, of pioneers who made their way, by land or by sea, towards a new place to settle down. The tradition of travelers' tales, the stories of distant areas are all familiar to many San Francisco Bay Area residents. In these stories, tales of the cities[14] destroyed by a natural disaster do not live only in the imagination, or in the hypothetical future. Referring to past earthquakes or destruction happening in others places is no coincidence: anticipating earthquakes is also imagining stories, some of which have already been partly heard and are known in the greater collective consciousness. Stories of cities that have been torn apart by large earthquakes still capture the attention of residents.

Whenever a major earthquake occurs anywhere around the globe, radio and television frequency signals relay the pressure waves and secondary waves across tectonic plates: information and pictures of the unfolding disasters are transmitted in almost real time and displayed on TV sets or computers.

In recent years, earthquakes happening in urban areas receive extensive media coverage in the San Francisco Bay Area. On January 12, 2010, an M7.0 earthquake, 25 km (15.5 miles) west of Port-au-Prince, killed approximately 316,000 people, making it the second deadliest earthquake ever recorded. On February 2, 2011, in Christchurch, New Zealand, a M6.3 earthquake killed 185 people; this was then followed by a shallow,[15] but still powerful M6.3 aftershock on June 13, 2011. On March 11, 2011, the M9.0 undersea megathrust earthquake[16] occurred off the coast of Japan, which is considered one of the five most powerful earthquakes ever recorded since record-keeping began in 1900; a wave[17]

[14]Reference is made here to those novels of Armistead Maupin that were set in San Francisco, which were first serialized and published in local newspapers before being turned into novels published between 1978 and 2014. The title was probably a reference to Charles Dickens' *Tales of Two Cities*, published in 1859.

[15]Earthquakes that originate within 60 km (40 miles) of the earth surface.

[16]Earthquakes that happen at a subduction zone, where one plate is forced underneath by another.

[17]"One person killed south of Crescent City, California and several boats and docks destroyed or damaged at Crescent City by a tsunami with a recorded wave height of 247 cm. Several houses, boats and docks destroyed or damaged at Santa Cruz, California; Brookings, Oregon; Hale'iwa, Kailua Kona and Kealakekua, Hawaii." http://earthquake.usgs.gov/learn/today/index.php?month=3&day=11&submit=View+Date.

generated by the earthquake hit the Pacific West Coast of the USA, killing 1 person. In Japan, the cumulative death toll of the earthquake and the following tsunami is estimated to have been 15,878 deaths with 2713 people still missing.[18] Despite the distances, and as the accounts of post Fukushima workshops have shown, these events have contributed to earthquake risk raise awareness of San Francisco Bay Area residents about the dangers of a similar disaster occurring locally.

The earthquake in Christchurch, New Zealand, is an other interesting example of the impact of a distant earthquake felt by residents of the San Francisco Bay Area. The *San Francisco Chronicle*, one of the main local newspapers in the Bay Area, covered the New Zealand earthquake in great details. Stories following both the immediate aftermath of the disaster and the reconstruction process, focused on the city's identity before and after the event. Spud Hilton, the *San Francisco Chronicle*'s special correspondent in Christchurch, was deeply moved by the spectacle of the torn, ghostly city and started to imagining how San Francisco would look like after such event.[19]

> Attempting to gauge what a comparable blow would look like in a U.S. city (based on size, population and landmarks), I envisioned San Francisco's financial district shuttered for a year, Grace Cathedral and St. Mary's both collapsed, Coit Tower in a pile of rubble and AT&T Park seemingly untouched, but red-tagged for demolition. And that wouldn't cover the thousands of homes in the suburbs (Hilton, 2012).

Personal identification with residents in Christchurch was strong; images of the 1906 earthquake, seen in numerous photographs immediately came to mind and Bay Area residents grieved the 185 casualties in New Zealand.

Changing the scale, disasters the 2010 Haiti earthquake received media coverage commensurate with the disaster itself: in a word, overwhelming. Just after the event, and before the first relief agencies arrived, reporters from all over the world gathered at the devastated airport of Port-au-Prince. In Haiti, citizen media response had already been on the

[18] In California, one death occurred as a result of this event, and several buildings were reported destroyed.

[19] "I had come to Christchurch to better understand how a series of quakes can turn a metropolis upside down" (Hilton, 2012).

alert when the earthquake first hit, and information circulated quickly through blogs, websites, social media (e.g., Twitter and Facebook), SMS, emails listservs,[20] and news reports. Information delivered through this network was far from simply being anecdotal ("Haiti: Earthquake!," 2010). For instance, The Haitian Blogger[21] started providing information immediately after the earthquake began. Sharing information as it emerged, line by line, she slowly drew a picture of the destruction, allowing her readers to accompany her in taking in the measure of the phenomenon:

> General Hospital in Port-au-Prince is down
>
> The Presidential Palace has collapsed – palè a kraze!
>
> President Renee Preval and his wife escaped injury.
>
> No one knows how many dead or injured.
>
> Report from an HLLN[22] member who was on the phone to Haiti told us that: "The road to Delmas 60 has collapsed down the mountain burying many homes. The people are screaming for help. Down the hill closer to Teleco there are a lot of UN troops on the street but that many of the roads are blocked with debrie [*sic.*] from collapsed homes…"
>
> […]
>
> People can only see dust – pousyè – dust, everywhere.
>
> Obama is sending in military/rescue troops and humanitarian aid.
>
> Phone lines that are working are: Haiti-tel and Voila.
>
> […]
>
> No landlines are working. no radio broadcast, no TV broadcast, no electricity…
>
> […]
>
> A terrible situation! Devastating. There's NEVER been an earthquake of this magnitude in Haiti. Major aftershocks happening…
>
> Nothing works, no one to assist anyone.
>
> No one knows where necessary personnel are.

[20] Just few hours after the disaster, emails began circulating, inquiring about members on email lists and discussion groups living in Port-au-Prince. One arrived in my email inbox that read: "With a lot of grief we were told that [X] had lost his 4 years old daughter [*sic*] and his mother-in-law during the earthquake. [Y], from who we had received an email a minute before the quake, was also killed in the collapse of this building."

[21] Chantal Laurent is a French-Haitian architect and an expert in emergency and reconstruction.

[22] Most probably, this referred to the Haitian Layer Leadership Network, but I have not been able to confirm this.

> [...]
> Someone in Haiti said to me "The area of Karrefour is destroyed...the population of Port au Prince has just been REDUCED, don't know by about how much. Everyone, rich and poor, build on the mountains, all the mountains are down! This is going back to the ground zero.
> This is catastrophic. Changes everything. ("Earthquake Rocks Haiti," 2010)

Lifelines, significant buildings, destruction, death, anxiety, data, and questions: the space of the city, the space of the everyday, was gone, replaced by a space of loss and death.[23] This blog post, as well as pictures of the destruction that soon followed, provided a dramatic dimension to the catastrophe, unfolding the new geography of Port-au-Prince for distant readers. "Because I had been there, when I saw those aerial shots I was sick to my stomach as I knew what it meant" [K.37], recalled one respondent. Facing the seemingly bottomless pit of loss and suffering,[24] people in the Bay Area reacted strongly. The emotional impact was fueled largely from testimonies and heart-breaking personal stories that quickly reached a considerable, yet disparate audience. "Right now my heart aches for Haiti, [...] my thoughts and prayers go out to the people of Haiti, and anyone with friends or family in Haiti," as one interviewee stated in the *San Francisco Chronicle* ("Haiti: Earthquake!," 2010). After the shock, compassion for the Haitians grew, and people sent not only prayers, but also money. In the USA, people's response to the disaster was concomitant with its nature, and $9 billion were collected for Haitian relief.[25]

Reaction to the Haiti disaster also went beyond the delegation of powers (financial and spiritual). People in the San Francisco Bay Area volunteered to work hands-on to help the Haitians. The responses took many forms. In Mountain View,[26] which sits in a southern section of the

[23]The expression "space of death" was used by Michael Taussig as his book title when writing on the event (Taussig, 2009).

[24]The 2010 Haitian earthquake is the second deadliest recorded earthquake of all time, after the 1556 Shaanxi earthquake, which occurred in China in 1556.

[25]In 2013, questions emerged about the lack of transparency regarding the spending of these funds (Ramachandran & Walz, 2013).

[26]The Google project was not the only crowdsourcing project that reached the USA. Others examples include a Boston group that allowed the geolocation of an emergency by SMS, and a group of Haitian expatriates that helped to translate data in French Creole into English and Spanish, which was then passed along to appropriate aid organizations.

Bay Area (locally known as the "the Peninsula"), a group of engineers working at the Google campus, developed "Google Persons Finder," an open-source project[27] based on the "People Finder" interface format setup after Hurricane Katrina in 2005. The application was launched three days after the Haiti earthquake in English, French, and Creole. A register of missing persons, the application was able to aggregate data from many outside sources working with the U.S. Department of State. Several weeks after the earthquake, the same team of Google engineers visited Haiti and subsequently created "Google Crisis Response."[28] These technical tools improved the circulation of information and data between first responders and beneficiaries, as well as amongst local communities. Creating connections was also very important to the Haitian community located in Oakland, California. Local newspapers relayed stories about the anxiety and despair of families trying to gather information about loved ones residing in Haiti:

> Longtime Oakland resident Antoine Bellot was at work when he heard the news of the quake, and hoped desperately that that his brother and sister in Port-au-Prince were not two of the untold thousands who were killed, crushed, or buried alive. He had just talked to his brother in Port-au-Prince on the phone an hour before the earthquake, but it would be days before he would hear news of their fate. "I couldn't sleep," Bellot said. "I was watching TV day and night, and I began to even see people that looked like my sister – but it wasn't my sister. It was heavy. I can't even explain it – it was a very painful situation. (Schoneker, 2010)

The fate of grieving families, those injured in the earthquake, and traumatized survivors, all with no resources to help them, deeply moved local Bay Area residents. To face this tragedy, the San Francisco Bay Area Haitian community got itself organized, and was soon supported by volunteers and donors from all around the area. The Haitian Action

Another company, Ushahidi, was key in crowd mapping. The initiative was encourage by FEMA: "On the third day, FEMA (Federal Emergency Management Agency) called us to say keep mapping no matter what people say – it's saving lives" (Morisy, 2011).

[27] http://google.org/personfinder/global/home.html.

[28] The Google person finder came under criticism after the Japanese earthquake and tsunami because of its total lack of protection of personal information.

Committee,[29] an Oakland-based organization founded in 1991, focused its efforts on fundraising to assist the search for missing people and to aid communication between victims and families.[30] Several street demonstrations, speeches, and concerts were organized in support of recovery efforts, which were relayed by newspapers, radio, and television. The compassion and sorrow for the people of Haiti was visible and materialized in the urban space, but unlike the aftermath of the Christchurch earthquake, residents' identification with the city itself, Port au Prince, did not occur. The history of Haiti (Farmer, 2011), the underdevelopment of the country and poor building codes did not seem to allow for a deeper identification with Haiti's capital city: "I don't think, if an earthquake happened here, it will be as complete a disaster as it was in Haiti, because of the style of most of our buildings, and because most of our buildings are relatively new and built to a fairly good codes" [A-M.2], noted a local San Francisco Bay Area resident.

However, Haiti earthquake was a moment when, as Susan Sontag has written, we truly regarded the pain of others (Sontag, 2004): watching the devastation was not only viewing the suffering of millions of victims but also, to a certain extent, sharing in their suffering. The personal and community initiatives that flourished in the aftermath of the event created a "window of attention," as disaster experts called it. Two years after the disaster, the emotional aspect of the earthquake could still be strongly heard during interviews: "It is painful to watch so many people suffering" [M.3], pointed out one expert who was still deeply moved by the memory of the images. Through television reports, the disaster had entered the realm of the viewer's experience, operating as a hybridization of their own experience of risk and disaster. As pointed out by another respondent, "When you talk about damage from earthquake it is not

[29]The Haiti Action Committee and other local groups organized events in the community to help the relief effort in Port-au-Prince. For example, in January 2010 the San Francisco Boys Choir performed at the Christ the Light Cathedral in Oakland, California, and Pierre Labossiere spoke at a fundraiser at the Black Dot Café. Additionally, the Haiti Action Committee organized a demonstration in downtown San Francisco (at Market and Powell Streets) to protest the U.S. military presence in Haiti.

[30]To date, approximately $5.2 billion has been pledged, amounting to about 80% of Haiti's entire GDP (Holden, 2011). In 2012, controversies began to emerge regarding how the money that was collected was being spent (Sontag, 2012). See also *Assistance Mortelle* (Mortal Assistance), a documentary motion picture broadcast on Arte on April 16, 2013 (Peck, 2013).

abstract. It became real and it's real people who you see, it is real people whom you are engaged with" [T.5].

Finally, the last major seismic event that shocked the San Francisco Bay Area residents and experts was the 2011 Japanese Tōhoku earthquake and tsunami, which occurred on March 11, 2011.[31] Throughout the years, California has developed "seismic complicity" with Japan, more than with any other country that sits on the shifting Pacific plates—also known as the "the Ring of Fire." For many decades, Japan has been known to be at the cutting edge of technology when it comes to earthquake safety, and has long been considered an example for California to follow. Consequently, when the Tōhoku earthquake and tsunami devastated the Fukushima Prefecture, the extent of the catastrophe caught Bay Area residents off guard (Fig. 4.2).

For many residents and experts Berkeley and Oakland, initial knowledge of the Japanese disaster came largely though emails and tweets exchanged within the diasporic Japanese community, as well as with those residing in Japan. Hitoshi, a San Francisco resident and business owner, recalled reading tweets about the event even before television coverage started: "I really got a sense that something amazing was happening across the ocean" (Yamada, 2012). To make sure that information was followed up by his readers, he used his blog to post updates and information that were subsequently shared thousands of times. "The calamity that shook Japan sent shock-waves across the Pacific, acutely felt in the Bay Area, inspiring action, charity and cooperation, opening a new chapter in a relationship bound by history, immigration, culture, commerce and time" was optimistically written in a report on philanthropic aid concerning the March 11 natural disaster (Yamada, 2012).

The Japanese Cultural and Community Center of Northern California, which had collected funds in the aftermath of the Kobe earthquake in 1995, was again chosen by the community to collect donations. The Japanese Cultural Center, situated in San Francisco's "Japan town" neighborhood, partnered with a local TV broadcast to launch a telethon. "With the advent of the Internet and social media, we were shocked at how successful we were initially," noted one member. A remarkable 48% of the $17.3 million culled from across the United States for Japan's reconstruction came from the San Francisco Bay Area alone. While, in

[31] As of March 2019.

March 17, 2011. The coverage of the Japan earthquake is live on every network channel. Images of the tsunami sweeping away houses and cars, filmed from a CNN helicopter, are taking most of the connected world by surprise. I, like millions of other viewers, am glued to my screen watching *CNN Live*. The cataclysmic dimension of what is happening is breathtaking. The unhindered and seemingly unstoppable brown wave is progressing with unbelievable velocity, literally washing away small cars – like toys - trying to escape a 40.5 Meters -133 Feet - tsunami wave. From bridges and certain other higher locations, people are watching, pinched faces and anxious eyes. This view of the wave rushing inland is nearly inconceivable. Questions, fueled with disbelief and indignation, emerge: Are they really safe watching a tsunami wave? Is that really happening right now? Can these people survive the wave of water and the debris? How far inland will this wave travel?[1]

While the wave continues its way through houses and open fields, what we are really watching is people dying, in their cars, in their houses, live on CNN. But that story, for now, is left untold. The CNN's correspondent diligently comments on the wave, and only the wave. Listening to his report, one could get the impression that not a single living being is there; the cars move by themselves; the houses are empty. This is just a wave ... an enormous brown wave, but still just a wave.

April 5, 2011. News reports are overwhelming. For some reasons, it seems that this catastrophe emotionally hits the U.S. harder than the Haitian earthquake the year before. Expert television and news reporters keep talking about the similarity between Japan's building codes and those of the U.S. For many, including myself, feeling safe from natural disasters is gone.

Fig. 4.2 Notes from research field notes, C. Mazel-Cabasse, 2011

the USA charity concerts were a success, training for first responders and community organizations was provided directly to Japan.[32] Tools developed during the Haitian earthquake one year earlier were improved: Ka-Ping Yee, Google's Person Finder software engineer, stated: "After the 2011 earthquake in Japan, photos of handwritten lists of people's names at the shelters were posted online, and volunteers across the Internet entered those names into Google Person Finder to help people reconnect to their loved ones" (Whitney, 2013). As during the Haitian earthquake, connections across the globe helping in the relief effort—from collecting data, gathering supplies to meet initial needs, to locating individuals—were made possible because of the physical mobilization of individuals and the use of tools.

[32] As happens in nearly all catastrophes of this magnitude, the question of the use of the funds collected becomes an important topic of discussion. We shall not address the question here, as we are focused on the "instauration" of earthquake risk in the San Francisco Bay Area.

But, soon enough, one disaster led to another—also devastating, but in a very different way. The tsunami triggered by the earthquake disabled the emergency generators required to cool the reactors of the Fukushima Daiichi Power Plant in Fukushima Province. This caused a partial meltdown in several units of the plant, followed by evidence of explosions and suspected releases of radiation. The news of a nuclear threat quickly raised serious concerns across the Pacific. Feelings went from compassion to anxiety when California appeared to be in the fallout zone of a nuclear cloud generated by the meltdown. Since 1986, Berkeley had proudly claimed itself a "nuclear free zone," and the concept of nuclear safety had remained one of the major concerns of longtime residents and citizens. After the earthquake, everywhere, customers in supermarkets, drivers in parking lots, parents with their children at playgrounds, people amongst their friends and families were sharing information about the potential danger of puddles of radioactive water,[33] and clouds hung over the Bay Area, instilling anxiety about the potential radio activity level of the food and soil.[34]

To address these concerns, on April 20, 2011, the UC Berkeley put together a panel discussion to evaluate the consequences of the Japanese earthquake.[35] That evening, several researchers shared their thoughts in front of a large audience. At a time when any proven consequences of the earthquake, tsunami, and nuclear meltdown were still very unclear, important discussions started around the question of nuclear safety. Panelists were sharing open-minded opinions about the risks of nuclear exposure in earthquake countries. During the discussion, a professor of nuclear sciences at UC Berkeley and a researcher at a private institute made it clear that nuclear technology should be more deeply tied into the democratic process. "We'll have to make our process clear enough for the public to understand," said the professor; "every country (*sic*) where the democratic system is working is lowering the importance of nuclear energy, all except France," added the researcher. The audience applauded, but remained skeptical regarding the capabilities of the Japanese authorities to address the disaster adequately.

[33] It rained for some days after the event.

[34] The "park bench" discussions, which can just as easily occur in local residents' kitchens, suddenly bring a considerable number of scientific references, but also put into play differing cultures.

[35] Cf. http://events.berkeley.edu/index.php/calendar/sn/ieas.html?event_ID=42584.

While experts in conference rooms were pleading for democracy, the case for science was far from being settled in other public spaces. Debates were raging as residents collected pieces of information from around the globe in many different languages.[36] Of course, not all of this information was reliable, and while a Radioactive Fallout Map started to circulate via social media,[37] it spurred a great deal of discussion and engendered a massive flow of comments across the Web, together with significant of misinformation and toxic posts.[38] But, online and on the ground, residents' and citizens' concern was real, and social media provided a platform for the dissemination of alternative information and independent research (Abe, 2013) at a time when official information was substantially lacking (Slater et al., 2012).

For several months following the Japanese event, the effects of the disaster created an important transformation in the awareness of seismic risks in the San Francisco Bay Area. Yet again, the possibility of a major earthquake became a not-so-distant reality.

The earthquakes in New Zealand, Haiti, and Japan triggered important questions and concerns about the potential consequences of a similar event occurring in the Bay Area, as well as the level of preparedness of local residents.[39] They reopened a Pandora's box of the potential

[36]"On March 22, France's nuclear safety institute started to release online bulletins in Japanese (Institut de Radioprotection et de Sûreté Nucléaire, 2011)" (Slater, Keiko, & Kindstrand, 2012).

[37]As Laura Beltz Imaoka (2013) notes, the use of this map, generated by the United States Nuclear Regulatory Commission as a indicative model, was also done without verification of the facts or any mention of independent, objective science.

[38]This map reprinted on millenarist and extremist blogs are similar one called Galactic News, which claims to "consider the future of our world," along with many other websites in the same vein.

[39]In addition to the earthquakes in New Zealand, Haiti, and Japan, Chile, Argentina, Loyalty Island, Southern Pakistan, the Fox Islands, Aleutian Islands, the Kermadec Islands region, Vanuatu, the Fiji region, Eastern Turkey, and the Eastern New Guinea region all experienced several earthquakes at the level of M7.0 or above. Virginia, a relatively earthquake-free state in the USA also experienced an M5.8 earthquake in 2011. On average, since 2010, there have been between 10 and 22 earthquakes at the level of M7.0 and above, and a world total between 1550 and 2494 earthquakes of any magnitude. However, despite speculation that the number of earthquakes may be increasing, the USGS chart shows a somewhat constant number of earthquakes for the 12 years since the turn of the millennium. The increasing number of registered earthquakes reflects the better coverage of various regions by seismic stations: there were 350 in 1931 in the USA; they are 8000 today. "According to long-term records (since about 1900), we expect about 17 major

Big One, although without any scientifically supported evidence of a connection between the different events.[40] Nonetheless, the series of small earthquakes that rattled the Bay Area the following autumn served as something like an echo chamber of the anxiety that had been growing for months.[41] As a consequence, when best sellers author Simon Winchester[42] stated in his column (Winchester, 2006), that San Francisco would soon experience a large magnitude earthquake,[43] Bay Area residents paid attention.

> It is as though the earth becomes like a great brass bell, which when struck by an enormous hammer blow on one side sets to vibrating and ringing from all over. Now there have been catastrophic events at three corners of the Pacific Plate—one in the northwest, on Friday; one in the

earthquakes (7.0–7.9) and one great earthquake (8.0 or above) in any given year" (USGS, 2013).

[40]These allegations were dismissed by researchers at the Berkeley Seismological Laboratory.

[41]"The first, and largest, was a 4.0 magnitude quake that hit just before 3 p.m. Thursday and was centered under South Berkeley, according to U.S. Geological Survey. The temblor was followed by three smaller earthquakes in the afternoon – a 1.0 magnitude quake at 3:02 p.m., a 1.8 magnitude quake at 3:09 p.m. and a 2.2 magnitude quake at 4:50 p.m. – and a 3.8 magnitude earthquake that hit at 8:16 p.m. that evening. According to the agency, no earthquake was centered in Berkeley on Friday. Just after midnight Saturday morning, a 2.8 magnitude earthquake shook the East Bay and was followed by two more smaller quakes within the hour that were also centered in Berkeley" (Karlamangla, 2011).

[42]Including: *A Crack in the Edge of the World: America and the Great California Earthquake of 1906* (Winchester, 2006)

[43]This statement does not contradict recent research. On its website, the USGS precisely states: "Along the Earth's plate boundaries, such as the San Andreas fault, segments exist where no large earthquakes have occurred for long intervals of time. Scientists term these segments "seismic gaps" and, in general, have been successful in forecasting the time when some of the seismic gaps will produce large earthquakes. Geologic studies show that over the past 1400 to 1500 years large earthquakes have occurred at about 150-year intervals on the southern San Andreas fault. As the last large earthquake on the southern San Andreas occurred in 1857, that section of the fault is considered a likely location for an earthquake within the next few decades. The San Francisco Bay area has a slightly lower potential for a great earthquake, as less than 100 years have passed since the great 1906 earthquake; however, moderate-sized, potentially damaging earthquakes could occur in this area at any time" (Cf. When Could the Next Large Earthquake Occur Along the San Andreas Fault? U.S. Geological Survey. Retrieved March 15, 2012 from http://pubs.usgs.gov/gip/earthq3/when.html).

> southwest, last month; one in the southeast, last year. That leaves just one corner unaffected—the northeast. And the fault line in the northeast of the Pacific Plate is the San Andreas Fault, underpinning the city of San Francisco. All of which makes the geological community very apprehensive. All know that the San Andreas Fault is due to rupture one day—it last did so in 1906, and strains have built beneath it to a barely tolerable level. To rupture again, with unimaginable consequences for the millions who live above it, some triggering event has to occur. Now three events have occurred that might all be regarded as triggering events. There are in consequence a lot of thoughtful people in the American West who are very nervous indeed—wondering, as they often must do, whether the consent that permits them to inhabit so pleasant a place might be about to be withdrawn, sooner than they have supposed. (Winchester, 2011)

The immediate consequence of such an alarmist story—even though scientifically unfounded—sparked particular anxiety amongst the residents of the East Bay. Media, again, played an important role in the transmission of information—and misinformation. Surfing on the wave, a website named "quakeprediction.com", which has no connection with the USGS, published even more alarming predictions. "[The warnings] make me nervous" (Wang, 2011), said a Berkeley resident to a local journalist. If the false information was relatively easy to debunk, the confusion reached a pinnacle when a professor from a local State University sent a mass email claiming that a Magnitude 6.0 earthquake had a 30% chance of happening within three weeks. USGS administration reacted promptly, reaffirming that no one could predict earthquakes. But it was already too late: the rumor continued to circulate for many days, and it took many newspaper articles and radio reports to calm the situation and expose the misinformation.

Experience of an earthquake is a mediated operation that creates moments of collective attention, shaping the space of earthquake risk the definition of risk and the subject living "at risk." Like the event itself, the stories reaching residents have left marks—hardly visible, yet still definite footprints—in the minds and memories of many.

In the information age, images of ruins and suffering reaching the Bay Area have, each time, caught residents by surprise. These events have opened up the possibility of a disaster, making the presence of an earthquake more real. Fear, anxiety, and compassion: for residents risk of earthquake exist through their attention to the other events in other parts of the globe.

4.2 Emotions That Connect

Emotions are part of the experience of living with and framing the risk of earthquake; they define the quality and the intensity of the relations between the everyday life of residents. They attach incline and repulse. They also connect residents and experts and allow for the emergence of particular form of expertise: one informed by acknowledging the role experience plays in the construction identities, as well as professional and residential practices.

4.2.1 Joking About the End of the World

On March 29, 2011, UC Berkeley held an event called "Earthquake, tsunamis and nuclear fallout: Is California at risk like Japan?" Faculty from the Berkeley Institute of the Environment, College of Letters and Science,[44] the Department of Earth and Planetary Science, the Berkeley Seismological Laboratory, and the Pacific Earthquake Engineering Research Center were each asked to make a five-minute presentation on scientifically informed reports of the Japanese situation. That night, the 150 seats in the auditorium quickly filled and the conference was then broadcast into the lobby for the overflow crowd still outside. The conference offered an opportunity for audience members to ask a question about the risks of earthquakes in the Bay Area and what actions should be taken. It also allowed scholars to discuss state-of-the-art developments in many fields related to earthquake sciences and safety. Some of the elements presented aroused strong reactions, which included surprise and concern, but also laughter.

Laughter erupted after a speaker explained how infrastructure built in accordance with probabilistic and determinist approaches may not resist earthquakes in worst-case scenarios. As he mimicked a sad face, the speaker added that, despite their best efforts, the underwater section of the Bay Area Rapid Transit (BART) commuter tunnel that allows trains to circulate between the East Bay and San Francisco, may not be entirely safe if a major earthquake were to hit. The audience laughed again. He then smiled and joked that he could lose his job after telling such an

[44]No misspelling here; "Letters" is plural, whereas "Science" is written as singular. This demonstrates an interesting perspective of UC Berkeley's conception of the sciences.

"inconvenient truth" (Guggenheim, 2006). With that, people laughed even more.

Of course, the fact the BART tunnel underneath Bay could crumble onto a train full of passengers is not particularly hilarious but, even so, people laughed. This is far from the first time laughter has been the reaction to disaster: making jokes about earthquakes is, in California, a common practice. In the introduction to his 1995 book, Robert A. Stallings quotes the following famous joke about the "ultimate California earthquake":

> Speaker: I just bought an ocean front property in Nevada.
> Friend: What are you talking about? Nevada isn't on the Ocean!
> Speaker: After the Big one it will be! (Stallings, 1995: 5)

This joke is one of most recurrent ones about the Big One, despite—or because of—the inaccuracy of the prophecy. Regional rivalry within the USA is another important comic impulse and is largely used to point to the specificities of the state of California (cultural, politic, and economic) but also, it seems, with special irony when it comes to earthquakes.

Another example of California's capacity to use humor when it comes to earthquakes occurred following an earthquake centered in Virginia on August 23, 2011, at 1:51 p.m. Eastern Standard Time. The earthquake, which was felt in 12 states, was the biggest that the East Coast had suffered since 1944. The event generated a considerable number of tweets and Facebook comments. But so much activity for a M5.8 earthquake was received in California as a manifest case of over-reaction, triggering sarcastic comments: "Hey, East Coast, the entire West Coast is mocking you right now," said one tweet, while another questioned, "Really all this excitement over a 5.8 quake??? Come on East Coast, we have those for breakfast out here!!!!" The breakfast metaphor went on with another tweet: "5.8? That's what us, Californians, use to stir our coffee with!" Similar comments continued throughout the day. "I haven't heard from anyone on the East Coast because they are probably still sitting under their kitchen tables," and "I wouldn't even wake up to a 5.8 if I was asleep" ("Post Quake, West Coast Teases East on Social Networks," 2011). These tweets received a great deal of attention, as they came from largely followed Twitter accounts, and were echoed throughout other online social media and the local press.

Irony is a question of distance, as another tweet showed: "Everyone calm down. If this is an earthquake on the east coast we're *supposed to* react ironically" [emphasis added] (Matthews, 2011). But irony is also a question of degree. If those on the West Coast wholeheartedly joked about the August 23 Virginia earthquake, most of them admitted to unleashing their satirical posts only after being sure that the earthquake did not cause serious damage. On the late-night shows that evening, hosts continued to joke about the event. David Letterman on his CBS *Late Show with David Letterman* program pursued the theme: "They felt the earthquake at Martha's Vineyard. It was so bad; President Obama nearly missed a putt." Interviewed by a journalist, Andrew Lakoff commented that, for California residents, "a perverse consequence of living with the ongoing specter of catastrophe, is this sense of pride" ("Jaded West Coast Chuckles Over East Coast Quake," 2011).

As was visible on the occasion of the Virginia quake, this irony and seeming insensitivity are also typical of the stories that California residents like to share about earthquakes. The sense of pride of being an earthquake survivor can sometime be overestimated, according to certain experts, such as this structural engineer: "People who have lived through the earthquake, they think of Loma Prieta as THE earthquake, whereas that was not the Big One. They'll say 'oh well… I survived the earthquake'" [J.8]. On the other hand, downsizing the event is also a good way to introduce some irony in the discourse: "I talked to my friend, [and for him], the biggest thing that happened during Loma Prieta was that he ran out of beer" [K.7], as one San Francisco resident, working for a Federal Disaster Management Organization explained me. Loma Prieta earthquake is also associated with leisure-time memories for baseball fans: "The only other earthquake I can think of then was Loma Prieta. I was in Davis at the time. We were watching the World Series,[45] and it went off the air. […] In Sacramento there were waves in the pool" [K.7]. The anecdotal aspect of the stories, their focus on facetious or shallow aspects of the event—which left an untold number of fatalities, casualties, and structural damage—as well as the casual tone used to describe the recollections keep the subject of earthquakes from being too dramatic.

[45]The earthquake happened during the third game of the 1989 Baseball World Series; as a result, the earthquake event was broadcasted live on national television. Coincidentally, the World Series featured the two major baseball teams in the San Francisco Bay Area, the Oakland Athletics (also known as the A's) and the San Francisco Giants.

4.2.2 *Dealing with Fear, Defining an Identity*

For much Californian, this distanciation with earthquake risks is an important element of the definition of their regional identity.[46] Respondents recognized that being a member of the greater Bay Area includes sharing the risks of earthquakes: "[There is] a certain 'woo-hoo' element, hard to explain, just part of the mentality of living here" [R.5]. This so-called "woo-hoo" mentality recovers a certain amount of detachment towards natural disaster, balanced with a strong attachment the Bay Area. "I feel I can live here for the rest of my life, and never get bored" [R.5], noted one respondent. Earthquakes are considered part of the experience of Bay Area life, something one needs to get used to and adjust to mentally. Longtime resident, writer, and victim of the 1991 Oakland Fire, R. explains:

> I grew up in California, so I'm used to earthquakes small and large. They just happen; I don't live in fear. I think people that move into the area and didn't grow up with [the earthquake] get freaked out at a small tremor. We have a small tremor, and we're like "Oh, it's an earthquake!" We move on and we don't think about it. [R.5]

This same type of reaction can be heard in other discussions as well: "You just thought we'd all be walking around living out of fear?," asked one person, expert in post-disaster reconstruction. "Is it any more or less dangerous than when you live in hurricane or flood country" [M.3], mentioned an other. Associated with childhood memories, earthquakes have been part of long term residents education: "There was an earthquake in the 5th grade, I remember the teacher standing there saying 'Oh, it's earthquake time everyone, get under your desks'" [K.7]. As R. pointed out, this "woo-hoo" [R.5] mentality does not materialize all at once: "In their first year in California, new residents get very worried about earthquakes, especially those who come from earthquake-free areas" [J.13]. Eventually, after several years, if they can bear it, they adjust and begin to behave like everybody else: exhibiting behaviors such as shrugging their shoulders and making jokes. By shaping a residential identity, this particular attachment and distanciation system works

[46]As quoted before: "Distanciation in general refers to the stepping back or distancing of the observer or reader from an object of scrutiny. The process of distanciation. It has both a spatial and an emotional side" (Harvey, 1997).

as a mark of belonging, which—as we saw in the jokes by and about Californians—draws a boundary line between "us" and "them." Those who stay and can bear to live with the risk versus those who cannot and choose to leave. One respondent, J., who works in a departmental building for the city of Berkeley, recalls:

> There was a big earthquake in Los Angeles in the '70s and friends of mine had moved into Los Angeles from Philadelphia. They had never experienced an earthquake before and this one was fairly bad. They moved back immediately and they would never come back to California, even to visit! [J.13]

For longtime residents, the unrealized possibility of an earthquake has transformed the risk from the category of "potential danger" to just an other narrative. In this perspective, not being prepared for an earthquake is considered an experience of material detachment and a way of beating the odds, explained R., who lost her house in the 1991 Oakland Fire:

> I know all the things that I should do in terms of documenting my possessions, taking a film of them and keeping it somewhere else. But I think everything is just stuff and the most important thing is your lives. So, even if I lost all my stuff again, it would be really, really bad, but I don't live in fear of it. [R.5]

Fear is often mentioned, mostly, only to be dismissed. If respondents agree they cannot live in a state of constant fear, some admit that the feeling still lays in their minds or guts: "Living in the Bay Area, anywhere in California, we've all survived a bunch of earthquakes. I think you live in constant fear of the Big One: 'is it going to come in my lifetime?'" [T.5], one respondent explained. Fear has different modes of existence: for some, the feeling can suddenly erupt, just as happened in the fall of 2011 but risk fear of earthquake is also the result of the work of worst-case scenarios evaluation, as defined by one seismologist:

> I can tell you one big fear, and there is no answer to it. If there's a major earthquake, and it happens in September or October, and a fire gets started, and it's one of those Santa Ana wind days, there's nothing I can do: I can't get out of here. And I don't dwell on it because there's nothing I can do. It's the big fear we all have because there are typically fires after earthquakes, usually not by gas, electrical. [...] We can't put out a big fire. [P.1]

When it comes to the circulation of information, the role of the media—this immense echo chamber—increases empathy and the feeling of vulnerability, building a space of emotions and concerns.

But when it comes to making residential choices, to find the safest mode of transportation, or to pick a school for their children, experts and non-experts living in the Bay Area alike face the same problems. One respondent, for instance, newly installed in the Bay Area, a manager in a humanitarian organization that provides emergency assistance, struggled to articulate the different levels of his interactions with the earthquake risk, which are simultaneously personal, and professional. What seems to connect this layered experience is the latent anxiety which, during my interview with him, clearly affected both his personal and professional life.

> I was hired to manage the volunteer program in San Francisco, and then, two years later, I was promoted to this position. So, I came into this job for managing volunteers – but with no experience in disaster or emergency response whatsoever. [...] The media are very quick to sensationalize the statistics: the number of people affected, how big it is. It really hit me when I first moved here, but it's even worse now. There are times, when my wife's away with my son – they go to L.A., or they go on a trip, and I go, "Okay, now it can happen. Go for it now, because I'm on my own." And I can handle this, but, if it happened right now, I mean, my son's at home with a nanny and another kid, and my wife's in Oakland. We'd have to get home, and it's just... it's scary! But, it just never ends, if you think like that. [...] I think the constant struggle for me personally (and professionally) is, I get the idea, I understand the general idea of how to respond. I can do the job. But, I'm not prepared in that I have not [had] a lot of experience doing it. [J.14]

Fear is a powerful fuel for both action and inaction. When asked about his motivation to work in the field of earthquake preparedness, one respondent answered just one word: "Fear" [J.14]. Fear can be contagious and can go beyond the boundaries of professional and personal identities. It complicates the emergence of knowledge as a "bold object" (Latour, 2007), of expert as an unconnected "knowing subject" (Mialet, 2012), questioning expertise without emotion, science without experience. Tracing fear in the experts' experience shows the entangled identities between the residential and professional dimensions of experience of living in an earthquake country.

The feeling of fear—which, of course, not all in either the expert community or the academic world agree on—generates a great deal of controversy among respondents. As the next quote illustrates, however, fear is not a unidirectional movement; rather, it is a confused process that can lead to very different positions. A quote, which comes from a risk-prevention specialist, moves progressively from the personal pronoun of "they" to the more communal "we" and marks an acknowledgement of the difficulty of keeping fear at a distance, of being only an unconnected rational expert. The quote begins with a reference to the fear of flying[47]:

> This is completely irrational when you look at it: you've got a situation where there's relatively few airplane crashes versus car crashes. People will complain and not wear seat belts in their car; and then, they'll just go "bananas" when there's just one airplane crash, just be completely irrational and fear of flying! And that kind of either inappropriate over-reaction or inappropriate under-reaction is a big problem when you're dealing with catastrophic events. In some cases, we over-react, like in the case of airplanes versus car crashes, or (in the case of earthquakes) we can under-react, and completely ignore things because they somehow feel scary. [J.8]

For many years, the public policies approach to prevention has used fear as an incentive to increase residents' disaster preparedness plans. However, using fear as a trigger for change has proven largely inefficient. As one expert working to improve NGOs' and community organizations' capacities to respond to disasters explained succinctly, "we are widely in love in this country with using fear and threat to accomplish social change" [A.-M.28]. Further, what may work at the individual level and in particular circumstances often doesn't equate to good results when applied to groups. Drawing on the failure of the traffic safety campaigns, the same respondent made the connection between the uses of fear, the form of power connected to it, and the spectrum of action expected:

> You would only use threat and fear under of a couple of circumstances. One of them is if you actually have the authority and the ability to fulfill that threat. That is why you know the campaign for seat belt is "*Click it or Ticket.*" You either click it or they will give you a ticket. It works because

[47] In reference to Air France Flight 447, which crashed June 1, 2009, killing all 216 passengers and 12 aircrew aboard.

> they can give you a ticket. They could not say "click it or else ..." and have nothing to back it up ... no one would matter, so it works because there is a threat. Fear works in political campaigns and is used all the time very successfully because they are not asking you to make a behavioral change, all they asking you to do is be afraid of the other guy, and go into that polling booth and vote against them. You don't have to change behavior, you don't have to shop differently, you don't have to do anything differently, you just have to be scared enough to push the other button. So, fear works very well in that sense, but fear has never work well for behavior modification. [A.-M.28]

Fear is not loved, to say the least. "It bothers me when there are some earth scientists whose goal is to scare people when they talk to the public" [K.11], explained an USGS scientist. As another expert made explicit, a consensus has been reached to "talk straight," in other words, "to tell [the people] the story but try not to scare the pants off them" [M.3]. Nonetheless, fear circulates.

Among respondents, the distribution of fear varied considerably and, despite the rather frightening description of the damage that could be caused by an earthquake, fear of a major earthquake is certainly not the most common feeling experienced by Bay Area residents. However, it is still present and experts have a hard time finding a good way to address it. Scientific probabilities offer a little comfort for some, but not for all, as this researcher explains when discussing his own wife's anxiety:

> I talk about earthquakes a lot. I see a lot about them and I'm quite comfortable with the idea of probabilities. I understand there's a certain probability of an earthquake on this fault and even if there is, there's a probability that the ground shaking won't be at that level and even if it is, there's only a certain probability that my house would be damaged and I'm very comfortable with that succession. For [my wife] there could just be a big earthquake that destroys our house because she's not a scientist or seismologist and doesn't understand the sequence as well or the probability. There's just a big destructive effect and she can't forget about it because I talk about it all the time. [R.9]

Of course, few people are married to an enthusiastic scientist and, to the great relief of many residents of the Bay Area, they don't have to think constantly about the risk of large-scale earthquake.

4.2.3 *Habit, Denial and the Un-extraordinary Existence of Earthquake Risk*

Because earthquakes do not happen all the time, the Big One neither monopolizes the past nor colonizes the future.[48] The relative distance of the danger, and the related feeling of safety that comes from it, plays an important role in the way in which people pay attention to the risk. The lack of attention to earthquakes can take different forms, which, when examined in detail, provide nuances in people's understanding of their vulnerability.

If big events trigger anxiety, this is not the everyday existence of earthquake in the Bay Area. Most of the time, earthquakes are a very discrete presence in people's lives; fault lines tend to stay dormant for months, if not for years. Most of the seismic movements on the Hayward and San Andreas Faults are small enough to go unnoticed: this diluted existence of the earthquake risk helps many forget the potential dangers. The "risk," as defined by experts, slowly becomes a danger to others, a distant eventuality that a number of earthquake experts living in the Bay Area, themselves, tend to forget.

Earthquake risk also competes with multiple other issues for the attention of individuals and institutions. The profusion of risks respondents face is often considered overwhelming—again, both for individuals and organizations—and this, in part, explains their difficulties in focusing their attention on "only" one topic. In 2010, a respondent was complaining of the lack of funding for disaster and risk management: "nobody is paying attention to disasters because they [are] all focused on joblessness and financial issues"[49] [M.3]. Even for those in the San Francisco Bay Area, the risk of earthquakes is sometimes de-emphasized in light of more pressing issues. One respondent, who actively works in earthquake preparedness and who has been very concerned by public buildings' safety during her career, was involved in the seismic retrofitting projects at her children's school. There, she had to face competing priorities and could only accept the situation. She recalled:

[48] The expression is borrowed from Beck (2009).

[49] Here, the respondent is referring to the unstable economic situation, which began with the 2008 financial crisis.

> I was on the Emergency Preparedness Committee, and we had the Emergency Preparedness Director for the Oakland schools come to a meeting. He looked us in the eyes and said: "In the Oakland school district, I am dealing with children who need to learn to get under their desks when gunfire is happening around them, and how to walk to school around drug-dealers" he said, "so, when you come to me with earthquakes, you have to understand that it's always going to be a low priority, because we have issues that are high-probability, high-consequence events." And you kind of think: "Okay…". [J.15]

Competing priorities may not be the only reason why people choose not to think about the earthquake. Looking at their past experiences, the history of the place, and the environment, residents draw up their personal risk maps, make disaster preparedness plans, which they do not always follow. As this structural engineer explains:

> The corollary that's easy to understand is health. My dad had diabetes. He had glaucoma, surgery to his legs: all the horrible things that diabetes can do [to] you… And he eventually died of the ramifications of diabetes. When I'm good, I remember that. When I'm on [a] staircase at the gym, I'm remembering that, when I'm eating …but then, there are times when that candy bar, that pasta just looks very, very good. Because those things are long-term, it's hard to remember. It was very fresh, right after he died, but […] we all have the incredible ability to deny risk. [J.15]

The recognition of an absence of focused thinking, of the inattention given to earthquakes, is often framed using the term of "denial".[50] This negation of the possibility of a disaster is part of the history of the Bay Area, as I have discussed in Chapter 1. It is also an important frame that helps in understanding how contemporary residents "deal with"—or, in this case, consciously or not, refuse to deal with—the risk of a major earthquake (Fig. 4.3).

As opposed to the denial that I described in Chapter 1, the deliberate lack of attention to earthquakes sits in the grey zone of knowledge:

[50] This is also discussed by Anthony Oliver-Smith "Disaster exposes the way in which people construct or frame their peril (including the denial of it), the way they perceive their environment and their subsistence, and the way they invent explanation, constitute their mortality and project their continuity and promise into the future" (Hoffman & Oliver-Smith, 2002: 6).

Talking about risk and disaster in an economically, socially, and culturally dynamic place like the San Francisco Bay Area in 2009 could almost seems like a lake of politeness toward a place that has so much to offer: vibrant city life, major universities, Silicon Valley, the spectacular landscape, and the temperate weather. Coming to the Bay Area from Europe with my dissertation project, I was surprised to discover that what I thought was a problem, or at least, could be problematized, was not even considered relevant by many respondents: *"Most people that I know don't think about it that much or they think about it and worry about it but I don't think it affects the way that they live their lives with that much,"* [M.3] as summarized by the first disaster expert living in San Francisco whom I met upon my arrival in the Bay Area. During interviews, the shorts and "closed" responses came back often: *"it is not something that I feel concern with,"* [M.3] and *"That does not affect my day-to-day thinking at all"* [G.6]. For many residents, the earthquake risk, blended with everyday environment and habits, seemed to be diluted by the presence of friends and family. *"No, I don't think about that at all, I've lived in California my whole life, so it's not even part of my thinking"* [G.6], answered one respondent rather sharply. *"It's my home, my friends are here. There's a certain embedded sense of stick-around,"* [R.5] described another when talking about her attachment to the place and why, despite having lost her home in 1991, she decided to stay in the Bay Area.

Fig. 4.3 Notes from research field notes, C. Cabasse, 2011

a zone where imagination, emotion, experience, and scientific knowledge are combined; a zone where habit, the repetition of movement or thoughts, is shaped.

Understanding how people live with risk is at the cost of a hybridization between forms of knowledge, as is made clear by a longtime expert complaining about the lack of understanding of her fellow residents: "They don't understand because it is about an event that hasn't happened yet, and they can't imagine that this {a large size earthquake} will happen to them" [M.3]. People do not fully comprehend the problem, not because they do not know, but because they cannot imagine. Experts themselves have difficulties articulating the complex relations between knowledge and experience:

> I think that human beings are fascinating. And, I'd do the same thing, though. [...] If you don't have the knowledge of something, you fill it in with personal experience. And so, it's: "My grandmother went through the 1906 earthquake, I went through the 1989 earthquake," and—despite the fact that they didn't experience a very severe earthquake ground-shaking—they supplant what should be knowledge, with experience. And they tell themselves, "I'm okay." [J.15]

Earthquake risk is a phenomenon of changing nature and existence. Earthquakes are defined as high risk, low probability events. The low probability, this relative distance to the potentiality of the event, makes a space for jokes and humor in relation to the potentiality of catastrophe. High risk, as we have seen, reluctantly opens the door for the potentiality of a space of loss.

Respondents have explained in a great deal of detail the nuances of this blurry territory and the ways in which experiences transform their personnal and professional identities. When confronted with the possibility of a large earthquake, residents, experts and scientists evaluate and compare the urgency and the potentiality of danger. They also assess their vulnerability and their capacity for resilience, acknowledging the complexity of the exercise like tightrope walkers tracing their path between science and experience. As one resignedly commented, "uncertainty opens the door to emotion. It's not always bad, but it could be better" [D.21].

The use of denial covers several meanings, several ways *not* to deal with the meaning and the consequences of living in an active seismic zone. Whether residents turn the question around and laugh, or whether they either accept or refuse to address their own fears, they utilize different strategies to address the risk. These strategies convey several meanings, which enter the—"blessed," as William James would have said (Latour, 2013)—grey zone of our awareness. Does this mean that risk itself is denied or not addressed? The answer is no. Instead, I argue the opposite. These indirect, non-scientifically based, seemingly futile, and messy ways to relate to the risk, speak to only that: earthquake risks are real and catastrophes possible.

4.3 Hybrid Knowledge

If everybody and everything is, to a greater or a lesser degree, at risk in the Bay Area, living with the threat of an earthquake is more an adaptive process than a static, predefined paradigm. Acknowledging this dimension opens up possibilities to go one step further. To grasp the passing, evanescent, reality-based earthquake threat, experts, residents, and expert-residents have to pay attention, to remember, to imagine, and to learn. They must define what their values are and to be able to evaluate the costs associated with those valuables. They have to listen to their emotions and make assumptions based on scientifically established statistics, on experience but also sometimes on chance. In this chapter, we shall see how this way of thinking establishes a new "normal," as defined as living with the threat of earthquake, which can shed light on specific, localized definitions of risk and knowledge while also stressing its limits.

4.3.1 *Knowing the Risk*

When it comes to earthquake risk, "knowing" is a complex operation. Using hierarchical categories, and comparing and discussing the risks, not only allows to envision the potential dreadful consequences of a large-scale event, but also repositions the risks in the time frame of an individual life span and experience. As discussed by a respondent living in San Francisco involved in risk prevention and the development of building codes:

> With earthquakes, they're so rare and extreme that to understand them, you have to think of them in the spectrum of everyday risks, monthly risks, and yearly risks. These all get compiled together and, most people, whether they articulate it or not, they're aware of that difference. You rarely find people that dumb that they don't understand risk in their daily life. [It] doesn't mean they always make the informed decisions, but they have an innate understanding of the rarity of things. [D.21]

The safety measures advocated to prevent major damages in the event of an earthquake concern the sphere of domestic life. A space where experts living in the Bay Area have to make the same day-to-day decisions as anybody else: choosing a house, a school for their children, or a transportation system to go to work. And, like anybody else, experts-residents also must make their decisions based on the amount of information available; evaluate a complex situations,[51] and imagine—if they can—the worst for their families. "I own a 1908 house that I upgraded. My house is retrofitted and very well built, on good soil. I am personally at less risk than other people" [M.3], pointed out one respondent, who also confessed:

> Post-Loma Prieta and after the fire, I was very conscious of making a family disaster plan, which we always have. Everybody always knew what it

[51] We will discuss these in greater detail in Chapter 5.

> was. My nanny thought I was out of my mind .... Everybody had to learn the rules: "This is what you do in case of an event, this is where you go." I had a very, very explicit family disaster plan, and I made sure everyone remembered it. [M.3]

An other respondent made a different, much more radical choice: "We bought a brand-new house because it's up to the modern standards. The number one way to be safe against earthquakes is being in a modern house" [R.9]. A third expert, interviewed in her office at UC Berkeley, justified her attitude:

> I should have a safety pack but I don't, which is silly given that I've lived through a fire and I know all the things that I should do in terms of documenting all my possessions, taking a film [*sic*] and keeping it somewhere else. I'm probably a little bit more aware of where all my photos are, and what I might grab in a fire, if I have to leave. I also think everything is just stuff and the most important thing is your life. So, even if I lose all my stuff again, it would be really, really, really bad, but I don't live in the fear of it. [G.6]

Experts deal with the likelihood of earthquake in different ways, as do other residents of the Bay Area. As a previously non-expert resident, now an expert in the field, recalled, the risk of an earthquake is not, and never has been, "a given"; rather, it has been progressively instaured by researches, practices, and attention: "I'm from Massachusetts and New York State. When I moved here, it was the 1970s; [the idea of a major earthquake] wasn't in anybody's awareness" [J.12]. This ongoing instauration was not a one-way street but, rather, a slow elaboration of the capacities needed to understand both earth science and resident behavior in a reflective way. Experts coming from the perspective of "hard science," learn at their own expense how to be not only rational subjects, but also beneficiaries of their own expertise. In the process, they learned to deal with, and even to respect, residents' understanding and practices: "If it does not make sense for the people to retrofit their home, then it does not make sense by scaring them into doing it" [D.21], summarized one respondent, a structural engineer by training who, here, echoed this common Bay Area sentiment.

Moving away from a discourse that is easily taken for granted regarding the lack of preparation and the irrationality of the residents

(Geschwind, 2001; Stallings, 1995), and also taking their distances with unfruitful attempts to detach irrational thinking from idealistic, "pure" scientific knowledge, these experts accept that several "frames can be considered rational yet lead to radically different solutions" (Von Winterfieldt, Roselund, & Kisuse, 2000, p. 35). As the previously quoted respondent noted, taking this perspective opens up considerable possibilities:

> We need to define what is rational by what people do, rather than decide what's rational and say that they're not being rational. They *are* the definition of rational, and therefore we have to rethink what rational is. [Emphasis added] [D.21]

This also changes preconceived narratives about people's relations to risk[52] and, in a broader sense, their understanding of individual and collective dynamics. As a respondent who has been working for 30 years in the field of hazard mitigation and long-term disaster recovery planning recalled that the process of defining the risk of earthquake was often full of surprises:

> We did a male survey of people in the mid-1990s. It was intended to find out why people would choose to structural retrofit in their homes, and as part of that, we wanted to see the correlation with whether or not people have done the Red Cross kind of things, like food, water, and first aid. And it turned out – as a side-line, because we also asked their age and income – that the more educated you are,[53] the less likely it was that you're going to have food, water, [and] first-aid training; and the less likely it was that you would have made the structural changes to retrofit your house, regardless of income! [*Laughs*] And we thought, "Okay ... Somehow, when people get a lot of education, they tend to have more blind faith that the utility companies are going to come through and they're going to have food and water. And [they think] they don't need to do this, because they know that their house is going to fall on the ground and therefore they're going to fix it. Whereas the other group, which was less well-educated, was convinced that it was going back to that basic survival training." We were

[52] In the literal sense, see, for instance, *Merchants of Doubt: How a Handful of Scientists Obscured the Truth on Issues from Tobacco Smoke to Global Warming* (Oreskes & Conway, 2011).

[53] Again, please note the change of the personal pronoun as she discusses this history.

> trying to hypothesize why this was going on, the basic survival training that they knew: that food and water were important on a day-to-day basis because they're having to deal with it weekly, as they did their budgeting. And therefore: "I need to make sure that I have set aside a little extra so that I will have food and water in case of any emergency, not just a disaster."

Listening to the people they interviewed, graciously accepting they were likely to be surprised, or even challenged, and reframing their hypothesis, Bay Area experts worked alongside the research taking place in the social sciences that favored and validated their own approaches.

For instance, despite the lack of extensive risk preparedness— at least, to the degree to which experts would have wished—when the Loma Prieta earthquake hit, research conducted by the UCLA Institute for Social Sciences Research showed that "there was little evidence that people panicked [during the earthquake]" (Bourque & Russell, 1994). More recently, studies have also tended to confirm that residents' awareness of earthquake risks has continually grown over the last several decades,[54] conducting experts to expand their understanding of the conceptual framework used by residents.[55]

Building on their own experience, experts of the Bay Area were also starting to redefine "normal"—which often for people not living in an earthquake country is a world without hazards. Understanding that the earthquake risk overlaps previous situations, though without connections,—such as science and experience, they framed the contours of a situated but moving norm of "living with earthquake" that never seems to reach a perfect and definitive conclusion.

> Everybody expects that if there is a major earthquake, things are not going to be as normal. So, it's okay if you take three days before going back to work. If you have to spend some money to patch up cracks and repaint, that's fine. Sometimes, engineers always show you the pictures of the damage because we can always do better and prevent that damage, but most of the time we should be looking at the pictures and saying, "So what? Is that acceptable or not?" [D.21]

[54]The first study, conducted in 2009 by Issues & Answers, a market-research industry group, by the Bay Area chapter of the American Red Cross.

[55]See Von Winterfieldt et al. (2000).

Of course, the following argumentation also changes the perspective of the risk itself:

> If you imagine an event and think you can recover, you don't need mitigation. You can tradeoff between mitigation, responding, and recovery. That decides how we plan mitigation, which ones we prioritize, why some make sense in some places but not others, why it makes sense for some organizations but not [for] others. The reason individuals don't do mitigation is because they understand they'll be able to recover. [D.21]

This line of reasoning is one that another respondent followed. Having experienced the 1989 Loma Prieta earthquake, she seriously considered moving away from California:

> And then I realized: no. If I move back to Kansas where my husband is from, there are tornados and that freaks me out even more. And there are hurricanes on the East Coast, which is where I'm from. And so, I realized I just have to prepare. And so that's when I went out and got my little backpacks, filled those, and did the things I needed to do, to make sure I would be as prepared as I could be. And then, there comes a point in time where there's nothing I can do about it. [R.5]

The limits of this non-coercive model surface, this mental exercise that relies only on personal understanding and imagination of the earthquake,—questioning the capacity for action of both residents and experts.

> You can present the information: "This is what your highest risk might be," but what would you think? It depends on what your experience is. If you've experienced an earthquake, you might think that's your highest risk, even if there is dry brush in your backyard and it's the middle of August and wind is blowing. And people who have been through firestorms and [have been] burned out of their house[s] and had [them] rebuilt, that's their number one risk, even if they're sitting on the earthquake fault. It's counterintuitive sometimes to realize your number one risk, so we try and present the information and that's pretty much all you can do because then the information is there: that's what it means, and this is what you can do about it. [D.21]

Knowledge is supported by the capacity to imagine the unthinkable, and to expect and accept the consequences of a large-scale earthquake. As a result of this hybridized risk, residents and experts accept—and come to expect—the possibility of destruction of buildings, infrastructure, and the ensuing economic loss. To a certain extent, some of them also accept the possibilities of large number of fatalities:

> [A good] analogy for a damaging earthquake is the 1995 Kobe earthquake; there were about 6,000 fatalities. That was on a fault directly beneath the city, [and] that's why it's a [good] analogy for an earthquake on the Hayward fault. [R.9]

Unfortunately, experts' also know many preventive actions cannot be accomplished pre-emptively. Often, experts have to recognize that their capacity to prevent damage are limited: earthquake can go way beyond, or be just very different from, anything they had planned for:

> I don't expect my house to collapse, maybe some windows, but I expect the electricity and power and water go out. We will have no money because the electricity will go down and affect money machines. I don't think we'll have 200,000 or more people killed, but I expect quite a bit of damage, it will take some time to bring back the infrastructure. There's a lot of old infrastructure here, the water systems under the streets, the sewer systems are all old. There is no real good way to go in and replace all that in advance. [P.1]

Instauring the earthquake is a mental exercise that allows experts to improve their knowledge about residential practices and residents' capacities to define the danger. This never-ending work-in-progress is continually renegotiated, moving the cursor of acceptability. When new building construction is planned, when a child is born, when a new scientific discovery is unearthed, when a new law is voted on, or when the time comes to choose a new house, all of the micro-events that had previously been balanced must be reprocessed.

The instauration of the risk is not the implementation of zero-risk safety. How could that be? Rather, it is the renegotiation of what is an acceptable level of threat that people can afford, knowing that what can be done may never be sufficient to cope with the extent of the damage and destruction. Instauring the risk in defining a new norm of what is acceptable when living with the threat of earthquake, endlessly rephrasing the question: "ask yourself *if* you have *a* risk, ask *what* you have *at* risk. Just the awareness is important" [emphasis added] [D.21]. In many ways, the inability to think *of* the danger frames the limits of this tightrope-walking mental exercise.

4.3.2 *Attending Uncertainty: About Fate, Chance, and the Metaphysical Dimension of the Risk*

Experts and non-experts alike articulate knowledge in a context of uncertainty. The dimensions of the "living with an earthquake" experience are never the same. They are embodied in an individual's life history, and within the larger framework of their values and concerns, and their ability to articulate imagination and knowledge. "Living with" is also a metaphysical experience where experts' and residents' actions meet their limits: a space where chance, God, luck (good or bad), and statistical odds can coexist.

"Luck," like the *deus ex machina* in Moliere's plays, appears with no apparent regard for circumstances. For seismologists, luck plays an important role in the combination of an earthquake's timing and location in regard to human activity. If an earthquake happens during a busy workday rush hour, the potential for a large number of casualties is heightened. However, if the earthquake occurs at a considerable distance from a densely built environment, the amount of damage and the number of casualties is not likely to be significant. Consequently, when looking at past events seismologists emphasize the role of luck when discussing assumptions regarding possible future events:

> We were very lucky [in 1989], because [the earthquake] happened on the day of this game between the Giants and the A's. Everybody stopped work and went home to watch. Normally at 5:00 in the evening, the freeways would be full of people, but they were empty. Instead of having 200 or 400 people killed in the collapse, it was only 40 or 50. We weren't the epicenter region, the earthquake did not happen in Oakland or SF, it happened somewhere else. [M.3]

If good luck plays an active role in a better ending, Fortuna, a capricious goddess, can also bring bad luck. The daily newspaper accounts describing the last moments of victims of the Cypress Freeway collapse in 1989 referred to the tragedy with the vocabulary of Fate and Destiny: "had the quake struck seconds earlier or later, Mrs. Marsden would scarcely have been affected, apart from picking up some fallen items at home" explained an obituary in *The New York Times*, dedicated to 11 of the 42 people who died that day (Reinhold, Navarro, & Rabinovitz, 1989).

Fate and Chance play an important role in the shaping of the personal dimensions of an event. But fate is not so easy to accept, as another respondent, a victim of the Oakland fire in 1991, explained: "what was also complicated in this situation here was [that] fate was pretty indiscriminate" [R.5]. For Oakland fire survivors, the difficulty of accepting Fate was deeply entwined with the collective response to the successive and unknown phases of the reconstruction:

> There would be six houses on a block burned, one house didn't burn. The person in that house was feeling very lucky and also had a lot of guilt for surviving; until the neighbors start building these big houses where everything's brand new. And suddenly you're looking around your old house... The same for the one house on the block that burned and everybody else's house was fine, that person would feel like a victim: "Why me?!" Then they build this nice house and their neighbors see it. There were a lot of waves of resentment. It was emotionally complicated. [R.5]

In the next respondent's case, the sense of belonging to a "common humanity" changes the understanding of fate, implying a fair acceptance of human experience at large, which includes—but is not only comprised of—natural disasters. For this former political activist, now in his eighties, who faced the anti-Communist McCarthyism witch hunts of the 1950s, natural disasters are considered less dramatic, or remembered as less traumatic, than matters that his family endured during the years of his political persecution. The fact that he lost his home and all of his belongings in the 1991 fire does not have the same weight in his memory as the decades of physical and psychological harassment, and the infringements of his and his family's rights:

> It's hard for me to think of risk in a conventional sense, because, even though I had personal experiences that involved fire and many houses burned down in 1991, including ours, somehow that never registered in my mind as any kind of unique experience. It was unpleasant, but I didn't feel personally victimized in any unique way. I regarded that kind of risk as something that all human beings face, one way or another. So, I didn't

> spend a lot of time bemoaning that kind of risk and that was true of my family, my wife, and the children. The only times in our lives where we felt uniquely at risk was the political experience. That's something of which the children were also aware, even though they were very young. [L.18]

However, in a place where "there are approximately 282 Evangelical churches, 93 mainline churches, 54 Roman Catholic churches, 27 uncategorized churches, 17 Orthodox churches, 28 other places of worship that either do not fit into the previous categories or could not be determined, 56 Buddhist temples, 17 synagogues, 5 Hindu temples, 5 mosques, 4 interfaith centers, 1 Shinto temple, and 1 Bahai temple as of December 31, 2011",[56] all within the city of San Francisco, Luck, Fate, and Chance are also deeply attached to larger spiritual dimensions.

In a poem in memory of one of the youngest victims of the 1991 fire, which was based on the eulogy given in a San Francisco synagogue, Hal Wingard wrote: "And those who must cry / Can never know why / It's they who must bear circumstance / No way to explain / To those with the pain / It's all only matter of chance."[57] As the prayer is translated into a poem, the words intended for a grieving devoted audience are hybridized with other form of metaphysic to give a larger dimension of the experience of the "living with", in a space of risk that overlaps a space of loss and death.

Earthquake, fire, their past occurrences, and their future potentialities are always situated within a larger network of actants to the list of which, Luck, Fate, Destiny, and deities need to be added. They represent the missing link between the known and the unknown, shaping the

[56] The map appears with following disclaimer: "This map is provided as a public service by YWAM San Francisco. The information provided was researched through public sites and is for informational purposes only. The research for this projected concluded on December 31, 2011 and is currently updated periodically, but may not be accurate beyond that date and is subject to change without notice. YWAM San Francisco is not responsible for misinformation, or misrepresentation that is presented here. This map is a work in progress, and we encourage participation, so please email us if you find any missing, wrongly stated, wrongly categorized, etc. information".

[57] Copyright 2009, Estate of Hal Wingard. In memory of Segal Livnah, who—on October 20, 1991—lost her life in the Oakland fire. November 17, 1991. Words, based on the eulogy delivered by Rabbi Martin Levin, Monday, November 4, at Temple Emanuel in San Francisco, completed November 15, 1991, on an US Air flight from San Diego to San Francisco.

dimension of experience but also tracing the limits of science as the only resource with which to face the threat.

Powerful actants, Luck, Fate, and Chance do not only emerge when disasters hit; rather, they stay hidden, invisible, until they are called again to shed light on the past or to support an optimistic version of the future. Having been exposed to the fire disaster, a respondent stated: "I'm hoping that I beat the odds and that it doesn't happen to somebody twice..."[58] [G.6]. In planning for the next disaster, Hope also becomes an important ally. The one that may be able to win Fortuna's favors—to gain time, for instance—as the next respondent, an expert in post-disaster reconstruction, explains: "You know, the hardest thing is to get people to do something and hope that you have enough time that you are actually able to accomplish something before another event happens" [M.3].

Here again, the strong presence of past events, the indirect experience of disaster can be felt: "We have built for earthquakes," added another expert, "and hopefully we'll not be like Haiti" [S.16].

4.4 Transformative Aspects of an Earthquake

The everyday, frequent association with earthquakes that have not yet taken place is an experience strong enough to hybridize knowledge and, simultaneously, to change the nature of expertise. For the expert community, living through the oblivious moments of everyday life in a seismic zone and sharing the common fate of a potential threat is a transformative experience.

> As a seismologist, I individually think of earthquakes from a purely scientific perspective. That obviously builds into understanding what the likely effects of earthquakes are. As an individual and regular person living in the Bay Area, I am interested to know the kind of very real impact an earthquake would mean for me. I think that's an important combination, a lot of seismologists are spread around the world working on earthquake hazards

[58] In full, she stated: "So maybe I'm hoping that I beat the odds and that it doesn't happen to somebody twice, it's really interesting that I know a lot more about what you should do and I haven't done any of it, just too busy or in denial. You think it can't happen to you again, maybe I feel I'm now immune" [G.6].

> wherever they are, but actually living in an earthquake zone forces you to combine the scientific aspect [with] the personal and societal aspects. [R.9]

Living with the risk of earthquakes, waiting and planning for the next Big One, allows experts to step back and add a layer of lived experience to their scientific knowledge. Thought of as an advantage, rather as a weakness, this new layer of experience is considered part of the earthquake as a risk.

> I don't know if basically it changed me, but I know that I have been in number of damages caused by earthquake shortly afterward. I find that those trips had a major effect on me, in term of considering how serious earthquake risks are, and their consequences. I think it took me to look at what the consequences are in society and the value that society has. [T.3]

As William James suggested in 1906, waiting for and experiencing disasters may transform individual psyches. Connecting not only science and concerns, but also the past and the future collective and individual experiences and various forms of knowledge, direct or indirect experience of earthquakes has a strong impact on human soul: "The tough thing about earthquakes is that you don't get warning. Maybe that does, or doesn't change our psyche about it" [S.16], reflected one respondent, a victim of the fire and, for many years, involved in disaster prevention for the city of Oakland. In *After the Quake* (Murakami, 2002), Murakami's characters live through what psychologists call a "post-traumatic experience," which unfolds in several steps. As psychoanalysis T. Rosbow develops "first, strangeness—the loss of the familiar; second, the past intruding into the present with the physical/emotional sense of being 'shoved;' and third, most importantly, the sense of randomness that follows in the wake of traumatic events, which wipe out our needed sense of predictability and order" (Rosbrow, 2012, p. 221).

In the Bay Area, many people were deeply shocked after the Loma Prieta Earthquake in a way similar to Rosbrow's description. As one respondent observed, "After the 1989 Loma Prieta earthquake, about three days after, I woke up in a sweat. Like, 'Oh my God, I have to get out of here!'" [R.5]. In the novel *After the Quake*, the description of the effects of an earthquake on the characters portrays the "mysterious and profound way" in which those changes operate (Rosbrow, 2012, p. 216). For some of the interviewees, the "symptoms," as they referred to them, continued for years to lay dormant but latent in the back of their minds. During one interview, a respondent who had been called to evaluate the damages of the 1989 earthquake, recalled:

> I didn't realize that I had a very mild case of PTSD, Post-Traumatic Stress Syndrome [*sic*], after Loma Prieta. We were running on adrenaline for months … I had a much higher level of … I don't know, I don't have it anymore. But I did have it. And I didn't recognize it. They did a little bit to help us, just a tiny … [*trails off*]. It wasn't like there was counseling or anything. They had a debriefing. […] It wasn't until I felt better [that I realized I had had PTSD]. You don't know how sick you are until you get better, right? I had a higher level of anxiety and some sleeplessness; the level of fear was a lot greater, and helplessness. You know, like, what can we do? [J.8]

The emotions provoked by the experience of the disaster are deeply anchored in people's narrative of their life history, as the recollections from fire victims displayed during the photo exhibition spoken of in Chapter 1 made clear. Often described as a trauma, the experience of disaster creates a discontinuity of normal life in individuals, which calls into existence something about the possibility of loss and death that was not present on this scale before the event. This new presence can be felt at different intensities and, if it is soothed by time, it can come back, brutally and without warning.

During an interview with a respondent who was recollecting the difficult process of selling the land where her house had once stood before the 1991 Oakland fire, she recalled the moment when she saw, for the first time, the newly built house completed by the buyers of her land. That day, she was deeply moved, in a positive way, by the sight of the property, inhabited again—a promise of continuity and happiness—on land that she had cherished for many years. She recalled that she had not expected to be caught by such strong feelings at the time; without noticing it, she remembered she started to cry at the moment she saw the new house. And as she was telling me her story, more than 20 years later, tears again were rolling down her cheeks, and she asked me to stop recording:

> It completely surprised me, as I'm surprised right now. I've been pretty bland this whole interview, it's like I'm so distant from this experience,

> I'm surprising me that describing this moment can still bring this out. I'm completely shocked by why I'm crying right now, but it must say something .... [G.6]

For earthquake experts in the Bay Area, most of whom have personally experienced a natural disaster, their unique understanding of the intensity of what such events can call into existence is the place where their practices take root. Even if, for some, the burden of this knowledge is deeply haunting.

> I was in Iran,[59] standing in front of a school where 600 children died. You know, it gives new meaning about your original question: "What is risk? and what do we do about it?" You can look at a building and say: "Typical, unreinforced masonry building in stone, floors not tied together, and during [the] shock, the floor has collapsed, that is an interesting mechanism...." The floor has collapsed. The problem was, it was 15 minutes before lunchtime and there were 600 children in there. [T.3]

To fight the feeling of helplessness aroused after catastrophes, and the feeling of helplessness that can also be present waiting in fear for a disaster that could come, experts of the Bay Area have decided to dedicate their time to prevention and rescue preparation. Finding their own resilience, they have discovered a way to avoid being victims of the risk of a great earthquake.

> I tell you, whenever there is a small one, I always think: "Is this it?" I think, simply because I know more about earthquakes, I just have to go to work then, I have to start to help my community put things back together again. I will be less a "victim" in an earthquake, than I will be part of the recovery. And, you know, maybe that's part of it too? There was a horribly helpless feeling after 9/11. What could you do? You can't go to New York and start pulling bodies out of the wreckage, because there are none. Whereas, after an earthquake, I have the ability to put on my hard hat, put on my boots, and go out and help people. It's a very different experience. [J.15]

[59] The Bam earthquake in 2003 was extremely destructive. Measuring M6.6, the earthquake that hit Iran killed 26,271 people and injured about 30,000 more.

Experts in the Bay Area may not want only to be moved when the tectonic plates move. Some recognize their strong attachment to earthquakes—an attachment that takes on, in their own words, the vocabulary of addiction and dependence:

> K. calls us "Earthquake Junkies"! That's what he calls us. It's the Earthquake Junkies. The people who are, let's say very concerned, conscious, obsessed, whatever you want to call it, you know! [J.8]

Here again, if humor is used as a favored way to describe mission—these experts have acknowledged that the motives for their actions is rooted in experience. Moved, or even driven, emotions these experts have defined the scientific contour of their practices. Evolving at the center of a network of attention are able to articulate the multiple dimensions of living with uncertainty without forgetting the messier, but also more complex, instauration of the phenomenon of earthquakes.

Earthquake preparedness is an instauration in the Souriaun sense and Earthquake Junkies are defined as much by the extent of their knowledge as by their attachment to make the Bay Area earthquake resilient. In the figure of the amateur, emotion and technic are not opposed; rather, they are interdependent. Chapter 5 will detail some of the technical aspects of earthquake prevention in the Bay Area. Inheritors of an empirical scientific tradition, we'll see how the experts/amateurs pursue the decade-long objective of building an earthquake resilient Bay Area. Concerned about details, able to articulate the different scales of intervention—from the size of a beam to the larger political impact of earthquake preparedness, earthquake experts emerge are complex figures for whom experiences inform and even drive the scientific practices.

References

Abe, Y. (2013). *Why safecast matters: A case study in collective risk assessment.* A STS forum on Fukushima. Berkeley. Retrieved from http://fukushimaforum.wordpress.com.

Anderson, B., & Harrison, P. (2010). *Taking-place: Non-representational theory and geography*. Surrey: Ashgate.

Atkinson, G., & Wald, D. (2007). "Did you feel it?" Intensity data: A surprisingly good measure of earthquake ground motion. *Seismological Research Letters, 78*(3), 362–368. Retrieved from http://earthquake.usgs.gov/research/pager/prodandref/AtkinsonWaldDYFI.pdf.

Beck, U. (2009). Critical theory of world risk society: A cosmopolitan vision. *Constellations, 16*(1), 3–22.

Berkeley Earthquake: Much Chatter About 3.9 Tremblor, but Little Damage. (2011, October 20). *L.A. Now*. Retrieved from http://latimesblogs.latimes.com/lanow/2011/10/berkeley-earthquake-much-chatter-little-damage-about-39-temblor.html.

Berkeley Earthquake Hits During "Shakeout" Drill. (2011, October 20). *Los Angeles Time*. Retrieved from http://latimesblogs.latimes.com/lanow/2011/10/berkeley-earthquake-hits-during-shakeout-drill.html.

Bossu, R., Gilles, S., Mazet-Roux, G., Roussel, F., Frobert, L., & Kamb, L. (2011). Flash sourcing, or rapid detection and characterization of earthquake effects through website traffic analysis. *Annals of Geophysics, 54*(6), 716–727. https://doi.org/10.4401/ag-5265.

Bourque, L., & Russell, L. (1994). *Experiences during and responses to the Loma Prieta earthquake*. Sacramento: California Seismic Safety Commission.

Chang, K.-K., Salmon, C., Lee, B., Choi, J., & Zeldes, G. (2010). The influence of contextual factors on the selection of news frames—A Cross-National approach to the news coverage of severe acute respiratory syndrome (SRAS). In G. Golan, T. Jonhson, & W. Wanta (Eds.), *International media communication in a global age* (p. 488). London: Routledge.

Davis, M. (1998). *Ecology of fear. Los Angeles and the imagination of disasters*. New York: Vintage Books.

Deep Under Berkeley. (2011). Retrieved from http://seismo.berkeley.edu/blog/seismoblog.php/2011/10/20/deep-under-berkeley.

Earthquake Rocks Haiti. (2010). Retrieved from http://thehaitianblogger.blogspot.com/2010/01/earthquake-70-to-73-rocks-haiti.html.

Farmer, P. (2011). *Haiti after the earthquake*. New York: PublicAffairs.

Foucault, M. (1970). *The order of things: An archaeology of the human sciences*. London and New York: Routledge Classics.

Geschwind, C.-H. (2001). *California earthquake, science, risk and the politic of hazard mitigation*. Baltimore and London: The Johns Hopkins University Press.

Guggenheim, D. (2006). *An inconvenient truth*. Los Angeles, USA: Paramount Classic.

Haiti: Earthquake! (2010). Retrieved from http://globalvoicesonline.org/2010/01/13/haiti-earthquake/.

Harper, D. (2010). *Creation myth of Northern Ohlone (Coastonan) Indian*.

Harvey, D. (1997). *Justice, nature and the geography of difference*. Oxford: Blackwell.

Hilton, S. (2012, February). Christchurch finding its way back from devastation. *San Francisco Chronicle*. Retrieved from http://www.sfgate.com/travel/article/Christchurch-finding-its-way-back-from-devastation-3252789.php.

Hoffman, S. M., & Oliver-Smith, A. (2002). Catastrophe and culture. In S. M. Hoffman & A. Oliver-Smith (Eds.), *The anthropology of disasters*. School of American Research Press, James Currey.

Holden. (2011). *How much money has been given and spent for Haiti earthquake relief? Putting the numbers in perspective*. Retrieved from How much money has been given and spent for Haiti earthquake relief? Putting the numbers in perspective.

Imaoka, L. (2013). Mediated maps: Geographic interoperability issues following global disasters. In *Society for Social studies of Sciences*. San Diego.

Jaded West Coast chuckles Over East Coast Quake. (2011). Retrieved from http://abclocal.go.com/kfsn/story?section=news/national_world&id=8323344.

James, W. (1906). *On some mental effects of the earthquake. The Youth's Companion*. Reprinted in James, H. Jr., 1911, *Memories and studies* (H. James, Jr.). Longmans, Green (Vol. June). New York.

Karlamangla, S. (2011). *Eight earthquakes hit Berkeley over past several days*.

Latour, B. (2007). *Changer de société, refaire de la sociologie* (Poche Scie). Paris: La Découverte.

Latour, B. (2013). *An inquiry into modes of existence: An anthropology of the moderns*. Cambridge, MA: Harvard University Press.

LittleJohn, D. (2011). *The big one. A story of San Francisco*. Durham: Strategic Book Group.

Livingston, A. (2012). *Excited subjects: William James and the politics of radical empiricism: Theory and event* (Vol. 15). Project MUSE. Retrieved from https://muse.jhu.edu/article/491201.

Margolin, M. (2003). *The Olhone Way*. Berkeley: Heyday.

Matthews, D. (2011). *The best earthquake tweets*. Retrieved from http://www.washingtonpost.com/blogs/wonkblog/post/top-16-earthquake-tweets/2011/08/23/gIQANgmJZJ_blog.html.

Mialet, H. (2012). *Hawking incorporated: Stephen Hawking and the anthropology of the knowing subject*. Chicago and London: University of Chicago Press.

Morisy, M. (2011). *A year later, lessons for the media from the Haiti earthquake response*.

Murakami, H. (2002). *After the quake*. New York: First Vintage International Edition.

Oreskes, N., & Conway, E. (2011). *Merchants of doubt: How a handful of scientists obscured the truth on issues from tobacco smoke to global warming* (Reprint edition). New York: Bloomsbury Press.

Peck, R. (2013). Assistance Mortelle/Fatal Assistance. Arte France (in co-production with), Canal Overseas Productions (with the participation of), Centre National de la Cinématographie et de l'Image Animée (participation), Radio Télévision Belge Francofone (RTBF) (in co-production with), Radio Télévision Sui.

Post Quake, West Coast Teases East on Social Networks. (2011). Retrieved from http://sanfrancisco.cbslocal.com/2011/08/23/post-quake-west-coast-teases-east-on-social-networks/.

Ramachandran, V., & Walz, J. (2013). Haiti's earthquake generated a $9bn response—Where did the money go. *The Guardian*. Retrieved from http://www.guardian.co.uk/global-development/poverty-matters/2013/jan/14/haiti-earthquake-where-did-money-go.

Reinhold, R., Navarro, M., & Rabinovitz, J. (1989). *The freeway dead: Portraits from Oakland—A special report. 11 whose lives ended as quake crushed I-880*. Retrieved from http://www.nytimes.com/1989/10/29/us/freeway-dead-portraits-oakland-special-report-11-whose-lives-ended-quake-crushed.html?pagewanted=all&src=pm.

Rosbrow, T. (2012). Murakami's after the quake—The writer as waking dreamer and trauma analyst. *Psychoanalytic Dialogues: The International Journal of Relational Perspectives, 22*(2), 215–227.

Sanders, R. (2011). *Third modest quake within week sends message: Be prepared!* Retrieved from http://newscenter.berkeley.edu/2011/10/27/third-modest-quake-within-week-sends-message-be-prepared/.

Schoneker, J. (2010, January 24). *After the quake, Oakland's Haitian community reaches out to help*. Oakland North. Retrieved from http://oaklandnorth.net/2010/01/24/after-the-quake-oaklands-haitian-community-reaches-out-to-help/.

Slater, D., Keiko, N., & Kindstrand, L. (2012). Social media, information and political activism in Japan's 3.11 crisis. *The Asia-Pacific Journal: Japan Focus, 10*(24).

Sontag, S. (2004). *Regarding the pain of others*. London: Picador.

Sontag, D. (2012). Rebuilding Haiti lags after billions in post-quake AID. *The New York Times*. Retrieved from http://www.nytimes.com/2012/12/24/world/americas/in-aiding-quake-battered-haiti-lofty-hopes-and-hard-truths.html?ref=deborahsontag&gwh=F8EE9544A4C3DF90794C-2960C75BDD13.

Stallings, R. A. (1995). *Promoting risk: Constructing the earthquake threat (social problems and social issues)*. New York: Aldine de Gruyter.

Taussig, M. (2009). Culture of terror—Space of death: Roger Casement's Putumayo report and the explanation of torture. *Comparative Studies in Society and History, 26*(3), 467. https://doi.org/10.1017/S0010417500011105.

Taylor, W. M. (2006). Urban disasters: Visualising the fall of cities and the forming of human values. *Journal of Architecture, 11*(5), 603–612.

USGS. (2011). US Geological Survey, Department of the Interior.

USGS. (2013). *Are Earthquakes really on the increase?*

Von Winterfieldt, D., Roselund, N., & Kisuse, A. (2000). *Framing earthquake retrofitting decisions: The case of hillside home in Los Angeles*. Pacific Earthquake

Engineering Research Center. Retrieved from http://peer.berkeley.edu/publications/peer_reports/reports_2000/0003.pdf.

Wald, D. (2001). *Did you feel it? Community-made earthquake shaking maps.*

Walter, F. (2008). *Catastrophes. Une histoire culturelle XVI-XXIème siècle.* Paris: Editions du Seuil.

Wang, A. (2011, October). Earthquake Predictions trigger anxiety, fear in Berkeley community. *The Daily Californian.*

Whitney, L. (2013). *Google Person Finder helping to find missing in Philippine floods.* C/Net.

Winchester, S. (2006). *The Crack in the edge of the world: America and the great California earthquake of 1906.* New York: Harper Perennial.

Winchester, S. (2011, March 13). The scariest earthquake is yet to come. *Newsweek Magazine.* Retrieved from http://www.thedailybeast.com/newsweek/2011/03/13/the-scariest-earthquake-is-yet-to-come.html.

Yamada, K. (2012). *The Bay Area's response to Japan's 3–11 catastrophe.* Retrieved from http://japancenter.sfsu.edu/sites/sites7.sfsu.edu.japancenter/files/uploads/HowHopeSpringsfromDisaster.pdf.

CHAPTER 5

The Case for Not Letting San Francisco Collapse

In the years since the 1989 earthquake and the 1991 fire, the contours of disasters have changed and have acquired an intensity that speaks to their long-term consequences. For those who lost their house or a loved one in the 1991 fire, the memories of the event can still be vivid, sometimes causing a shortening of breath or bringing on tears that were once thought to have dried. The ramifications of the event still need to be traced and, in the process, the reticular chain that we are following will tell us about the constructions of space and risk in the Bay Area.

After the 2011 Japanese Tōhoku earthquake, Simon Winchester stated in a *Newsweek* column entitled "The Scariest Earthquake Is Yet to Come" (Winchester, 2011) that, after the series of earthquakes that hit the "Pacific Ring of Fire," San Francisco should—sooner, rather than later—expect a large-magnitude earthquake along the San Andreas Fault.[1]

[1] This statement does not contradict recent research. On its website, the USGS states: "Along the Earth's plate boundaries, such as the San Andreas fault, segments exist where no large earthquakes have occurred for long intervals of time. Scientists term these segments 'seismic gaps' and, in general, have been successful in forecasting the time when some of the seismic gaps will produce large earthquakes. Geologic studies show that over the past 1,400 to 1,500 years large earthquakes have occurred at about 150-year intervals on the southern San Andreas Fault. As the last large earthquake on the southern San Andreas occurred in 1857, that section of the fault is considered a likely location for an earthquake within the next few decades. The San Francisco Bay area has a slightly lower potential for a great earthquake, as less than 100 years have passed since the great 1906 earthquake; however, moderate-sized, potentially damaging earthquakes could occur in this area at any time" (U.S. Geological Survey, 2016)

C. Mazel-Cabasse, *Waiting for the Big One*,
https://doi.org/10.1007/978-3-030-15289-5_5

The statistics on earthquakes are well-known and Winchester's claims are only one instance of the most recent statements on the topic, over which much ink has been spilled. The claims can be summarized as follows: residents of the Bay Area have settled, and continue to do so, in a location where they should not be in the first place (Reisner, 2003), earthquake risk experts seem unable to relay their information to the general public (Geschwind, 2001).

How can such an important message as "be prepared" fail to reach its audience so badly? As Stallings underlines it, "With the possibilities of negative consequences from a future catastrophic earthquake seemingly as great as those from street crimes, epidemics, and other publicly recognized risks of the day, why has the earthquake threat attracted so little public attention and concern?" (Stallings, 1995: 193). In the brochure entitled "Putting Down Roots in Earthquake Country,"[2] the USGS provides a map indicating the possible shaking levels around the San Francisco Bay Area. The brochure specifies that soft soils, unlike bedrock, have the tendency to amplify and prolong shaking. It then goes on to say that most of the built urban areas of the greater San Francisco Bay have been constructed on these exact, dangerous soft soils.

5.1 The Tip of the Iceberg: Big One Scenarios

Earthquake risks are made visible by maps and scenarios that circle, define, and evaluate the potential for disruption, destruction, and death. As plays, sketches, and scripts, scientific scenarios assemble actants and look at the ways they interact with one another. They give "life" to a phenomenon that has not yet happened, but once instaured, they also become actants on their own, calling for actions and policies. Scenarios are the results of bringing together years of collective, multidisciplinary research that need to be articulated in a large enterprise of sense making. In summarizing their 30 years of research and predictions in this field, the group of scientists responsible for the development of seismic maps explicitly discuss interconnections and interactions:

> Drawing on new data and new methodologies, we have concluded that there is a 0.62 probability (i.e., a 62% probability) of a major, damaging

[2] Edited by the Southern California Earthquake Center, First Edition, 1995; Review Edition, 2003: 32 pages.

> earthquake striking the greater San Francisco Bay Region (SFBR) over the next 30 years (2002–2031). Such earthquakes are most likely to occur on seven main fault systems identified in this study, but may also occur on faults that were not characterized as part of the study (i.e., in the "background") […]. Our results come from a comprehensive analysis lead by the USGS and involving input from a broad group of geologists, seismologists, and other earth scientists representing government, academia and the private sector. The results of this study are appropriate for use in estimating seismic hazard in the SFBR, and estimating the intensity of ground shaking expected for specified "scenario" earthquakes. In addition, they provide a basis for calculating earthquake insurance premiums, planning and prioritizing expenditures for seismic upgrades of structures, and developing building codes. (Working Group on California Earthquake Probabilities, 2003)

Earthquake science in the twenty-first century is de facto hybrid, combining years of multi dimensional data analysis. To aggregate these variable elements, scenarios focus on precise research questions: the modelization of economic losses (Grossi & Zoback, 2009; Kircher, Seligson, Bouabid, & Morrow, 2006), the readiness of infrastructures and lifelines (Maffei, 2010), building constructions (Poland, 2009), the recovery capacity of local governments (Min & Perkins, 2008), or global prevention policies (Poland, 2009). All of these scenarios foresee the potentially devastating impact an earthquake can have on water supplies (e.g., the main water pipes cross the Hayward Fault),[3] as well as electric and gas distribution.

5.1.1 Grasping the Earthquake Phenomenon

Scientists and experts now agree that earthquakes in the Bay Area occur primarily because of the accumulation of "strain energy" caused by northwestward motion of the Pacific Plate away from the North American Plate, which moves in a southwestwardly direction towards the Mid-Atlantic Ridge.

[3]The Hetch Hetchy system is complex gravity-driven network of pipes, dams, reservoirs and tunnels that was completed in 1934 between Yosemite Valley and San Francisco. The water system collects water from the Tuolumne River and delivers 260 million gallons (or 984,207 cubic meters) of water per day to consumers in the Bay Area. Because of its age, the system is considered very vulnerable and liable to service disruption.

However, and despite decades of research, estimating the earthquake risk for the Bay Area remains difficult. Combining approaches—empirical laboratory experiments, observation and modelisation—scientists have tried to explain the presence of "earthquake clusters": the high frequency of earthquakes followed by prolonged periods of little activity. As one group of researchers stated:

> Between 1836 and 1911, moderate earthquakes were common in Northern California, and shocks of magnitude 6.5 to 7 occurred every 10 to 15 years. However, from 1911 to 1979, no temblors of even moderate magnitude occurred. In 1979, a new period of high earthquake activity, similar to the period prior to 1911, appears to have begun. The magnitude and year of each quake are shown. (Some magnitudes are recent updates based on the latest U.S. Geological Survey data). (Thatcher, Ward, Wald, Hendley, & Stauffer, 2001)

With sufficient data, it is possible to give a satisfactory representation of earthquake clusters. But charting the absence of earthquakes during the last several decades has tended to worry scientists: even if very large earthquakes are rare, moderate earthquakes, due to the build-up of strain energy, still have potential for a great amount of destruction.

As explained in one USGS report that was presented to evaluate the probability of earthquakes in the San Francisco region:

> Our analysis suggests a 30 years probability of an earthquake M7.5 or larger striking the region is only 0. 10 (0.02 to 0.20). Only the San Andreas and San Gregorio faults, both lying west of San Francisco Bay, have sufficient length to generate such a large event. [...] We estimated the probability of a moderate earthquake (M6.0 to M6.7) over the next 30 years to be at least 0.80 (at least four times as likely to happen as not). As the recent past has demonstrated, earthquakes of this magnitude and smaller can produce significant damage over localized areas. (Working Group on California Earthquake Probabilities, 2003)

Because of the density of the built environment in the San Francisco East Bay, probability estimations have been developed for both the San Andreas Fault and the Hayward Fault.

The last major earthquake on Hayward Fault was in 1868, with an established recurrence pattern of 140 years. In 2008, experts estimated that the Hayward-Rodgers Creek Fault system had a probability of rupturing of nearly 31%, and that the probability of an earthquake measuring M.6.7 or more within the next 30 years was estimated to be 63%.

The other major fault line that causes a great deal of concern is the San Andreas Fault, which has an estimated risk of rupturing at 21%. Previous major movements of the San Andreas Fault were responsible for the M.7.8 1906 earthquake and the M.6.9 Loma Prieta earthquake in 1989. Smaller faults, including the Calaveras Fault in the East Bay and the San Gregorio Fault along the San Francisco Peninsula have an average 6–7% probability of striking in the next 30 years.

The possible consequences of earthquake risks are difficult to grasp because they happen in heavily built-up metropolitan areas. If many scenarios use historical earthquakes as starting points from which to evaluate potential damage (Fehr, 2006; Heller, 2006; Kircher et al., 2006; Perkins, Chakos, Olson, Tobin, & Turner, 2006; Society, 2006), they also take into consideration the fact that the Bay Area's population is now 10 times greater than it was in 1906, and the total property value is currently estimated to be at least 500 times greater.

As at 2013, the population of the San Francisco Bay Area was estimated at 7.4 million people living in 2.7 million households (ABAG, 2013a). The regional economy in 2009 was estimated to be $300 billion.[4]

The San Francisco Bay Area is also a major industrial hub in the west coast of the United States. It hosts six refineries, and large roads crossing the faults take an estimated 250,000–300,000 commuters across the Bay Bridge every day (ABAG, 2013a; Chakos, 2011; Hinman & Hutchinson, 2005; Poland, 2009) and, of course, one of the largest center for innovation and technology: the Sillicon Valley. In this large metropolitan area, individuals tend to commute a lot. Large motorways, suspended highways and bridges connect the main cities, crossing the faults lines at multiple points of contact. The major commuter train system, the BART, crosses the fault system, using tunnels including one built on the floor of the San Francisco Bay (Grossi & Zoback, 2010). Finally, some of the most sensitive industrial and scientific installations in the region are also located very close to the faults system.

[4]It worth noting that most of the reports evaluating the risk of earthquakes, including not only those in the Bay Area, focus mainly on the economic data to picture what is commonly called "socioeconomic." Figures include growth, employment and unemployment, value of retail loss and properties, and other estimated losses.

> The high-tech industry drives employment in the South Bay, while the University of California and two national laboratories drive employment in the East Bay. In the North Bay, tourism, agriculture, and distribution and manufacturing dominate employment. The Peninsula receives spillover from San Francisco and the South Bay. Its economy is largely high tech and biotech. Major employers on the Peninsula include Oracle, Stanford University, and United Airlines (due to San Francisco International Airport). (Perkins, 2005)

For these reasons, many scenarios show that "of all American cities, San Francisco is probably the most vulnerable to catastrophic disaster" (Paxton, 2004). In addition:

> Our housing stock is the oldest in the West. Much of the city was built on sand, landfill, and other non-compacted soil, which is subject to liquefaction. And with 75 percent of the rental housing stock covered by rent control, much of the housing stock is poorly maintained. All of these factors exacerbate the dangers inherent to being precariously straddled between the San Andreas and Hayward faults. (Paxton, 2004)

Conducting meticulous evaluations of San Francisco buildings, researchers have estimated that between 7000 and 10,000 commercial buildings could be destroyed in the event of a major earthquake, and transportation, electricity, water, and the food system would be either partially or totally interrupted (ABAG, 2013b; Bonowitz, 2009; Tierney, 2000).

The local infrastructure is indeed very fragile and reports have consistently shown that the economic fallout and casualties of a significant earthquake could over-stretch the capacities of local and federal government responses (Perkins, 2005; Poland, 2009). "It would result in $260 billions of damages to residential and commercial exposures, of which $50 to $80 billion would be covered by property and workers compensation insurers" showed one study (Bradford, 2006; Grossi & Zoback, 2009) while other pointed to the devastating figure of 6000 fatalities[5] (Maffei, 2010; Perkins et al., 2006).

In the case of the Hayward Fault, calculations have been made according to two possible scenarios: either a rupture of the entire fault, or a partial rupture of the southern segment. According to different

[5] Estimation based on the number of fatalities caused by the 6.8 (Moment Magnitude Scale) Kobe earthquake in 1995.

scenarios, the total economic loss could be around \$95 billion and \$190 billion, impacting as much as 3.4 million jobs (Hutchings, Mieler, & Brechwald, 2013).

These figures would make any resident dizzy, and they are certainly not uncontested and undiscussed. But they are expert's best estimate of what could happen in case of a large earthquake. Looking at the condition of production of these astonishing figures, one realised that they are the result of years of cross analysis and debates from groups of experts and scientists. Operating in a process of constant translation, changing scales, and aggregating material, these scientists produce reference documents that define the earthquake risk according to the state of knowledge at particular moments in time. One of these important documents is the USGS Seismic Hazard Map.

5.1.2 *The Hazard Map: Getting Concerned About Earthquakes*

Scenarios are not the final step of the construction of earthquake risk; rather, they are just the starting point. The USGS Seismic Hazard Map is a good example of a progressive instauration of risk. The Seismic Hazard Map sits at the nexus of geophysical data relating to the movement of the Earth's crust, and the definition of seismic risk. It makes visible information collected by various scientific methods: systematic collection, standardization, and multiple verifications. This capacity for data translation and communication makes it a pivotal actant of risk definition that feeds policy planning at the local, state and federal level. Its final production brings together a long chain of facts and figures, tools, funding agencies, political will, and organizational cooperation that comprise the familiar path of major scientific research (Lynch, 2012).

The Seismic Hazard Map is also included in recommendations by the National Earthquake Hazard Reduction Program[6] (NEHRP). As such, it

[6] "The activities of the Program shall be designed to: (A)[…] research and develop effective methods, tools, and technologies to reduce the risk posed by earthquakes to the built environment, especially to lessen the risk to existing structures and lifelines; (B) improve the understanding of earthquakes and their effects on households, businesses, communities, buildings, structures, and lifelines, through interdisciplinary and multi-disciplinary research that involves engineering, natural sciences, and social sciences; and (C) facilitate the adoption of earthquake risk reduction measures by households, businesses, communities, local, state, and federal governments, national standards and model building code organizations, architects and engineers, building owners, and others with a role in planning

plays a significant part in the creation of building codes[7] by the Building Seismic Safety Council[8] (BSSC), and the retrofitting guidelines designated by the Federal Emergency Management Agency (FEMA). It is also used to evaluate the construction of public infrastructure projects and it has been embedded in the lay out of several federal and state programs. The map is also an important source of information for the financial sector: the California Earthquake Authority (CEA) uses it to define premiums for state insurance programs and financial companies, such as those who oversee pension funds, in order to evaluate risk in their portfolios.

Overall it provides "the scientific basis of seismic provision in building codes enacted throughout the United States to prevent loss of life and limit damage during large earthquakes" (Filson, McCarthy, Ellsworth, & Zorback 2003: 2).

The 2008 version of the map is based on the successive accumulation of data, corrections, and new measurements undertaken since its first publication in 1976. It incorporates new elements: the probability of larger earthquakes based on the long-term history of earthquake magnitudes in various regions and changes the model of ground acceleration.[9]

The National Seismic Hazard Mapping Project (NSHMP) works to combine and evaluate actions of the already intricate associations of human and non-human actants: faults, earthquake ground shaking, geodesy[10] and seismicity,[11] seismic soils, near-surface conditions, energy

for disasters and planning, constructing, retrofitting, and insuring buildings, structures, and lifelines through: (i) grants, contracts, cooperative agreements, and technical assistance; (ii) development of standards, guidelines, voluntary consensus standards, and other design guidance for earthquake hazards risk reduction for buildings, structures, and lifelines; (iii) outreach and information dissemination to communities on location-specific earthquake hazards and methods to reduce the risks from those hazards; and (iv) development and maintenance of a repository of information, including technical data, on seismic risk and hazards reduction" (112th Congress 1st Session, S.646 To reauthorize Federal Natural Hazards Reduction Programs and for others purposes, In the Senate of the United States, March 17, 2011).

[7] The building code has been adopted by 37 states, including California.

[8] The BSSC, established by the National Institute of Building Sciences, develops and promotes building earthquake mitigation regulatory provisions for the whole nation.

[9] The 2008 version includes new data on the subject of fault slip rates, paleoseismologic data from fault trenching studies, earthquake catalogs, and strong-motion recordings from global earthquakes.

[10] Geodesy is a branch of Earth sciences that deals with the question of measuring the Earth. Amongst other work, it focuses on crustal motion.

[11] Seismicity refers to the geographic and historical distribution of earthquakes.

attenuation and paleoseismology, to gain a better understanding of the physics involved in the Earth's crust. It used most recent technologies extensively, such as global positioning systems (GPS) and visualization. The production of the map is a process comprising several steps, the first of which is that NSHMP experts gather to "discuss progress on the map, input data and procedures used in the process" (Petersen et al., 2011: 9) (Fig. 5.1).

Following an initial meeting by NSHMP experts, the Working Group on California Earthquake Probabilities (WGCEP), the California Geological Survey (CGS), and the Southern California Earthquake Center (SCEC) together determine the most accurate methodology for earthquake forecasting (Fig. 5.2). It is revised approximately every six years by NSHMP, and always being updated with new developments in seismic research.[12]

The map works like a transmitter, opening a window of continuity, and movement[13] between the realm of everyday life and the realm of possible events and scenarios. The next section explores the organisation of the collaborative work at the origin of the development of the map. The next sections will then examine other dimensions and ramifications of these productions, from the building of extensive and diverse tool box to the development of public policies.

5.1.3 An "Earthquake Junkie" at Work

Stallings (1995) has defined them as claim-makers: scientists and experts writing up scenarios and reports, drawing maps, participating in panels, sitting on committees, being involved in congressional hearings, and being interviewed by nationals news organizations.[14] Geschwind (2001) has looked at them through the lens of their specific organizations as they work to consolidate their capacity to transform state regulations

[12]The Next Generation Attenuation Relation (NGA) was implemented by the Pacific Earthquake Engineering Research Center (PEER) and conducted ground breaking (so to speak) discoveries in ground motion.

[13]This movement is an also active connection between the forms of knowledge, methodological epistemologies, and disciplinary field research.

[14]According to his research, they are composed of academics (28.05%), the federal government (33.84%), state and local governments (10.37%), international agencies (2.44%), the non-profit sector (12.50%), the private sector (10.67%), and politicians (2.13%).

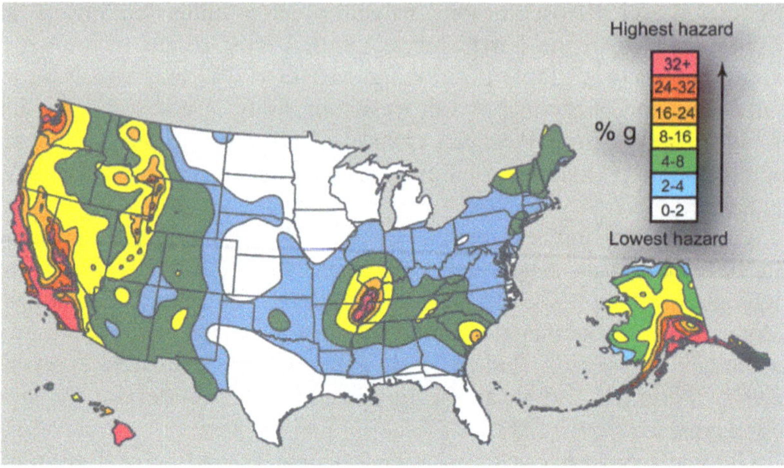

Fig. 5.1 USGS 2008 National Seismic Hazard Map (Credit: US Geological Survey Department of the Interior/USGS—U.S. Geological Survey)

Example of modifications to the Californian section of the Map:[1]

1. Revised earthquake catalog and accounted for magnitude round off and uncertainty.
2. Constrained model to fit within two sigmas historical or observed seismicity rates (suggested by WGCEP—Science Review Panel):
    2.1. Reduced moment rate on faults by 10 percent to account for aftershocks, foreshocks, after slip, and smaller earthquakes;
    2.2. Reduced earthquakes M6.5 in smoothed gridded seismicity to one-third the rate to account for earthquakes already modeled on faults (generally not applied outside California);
    2.3. Implemented a branch of Gutenberg-Richter Model with b=0, which is consistent with modeling of several of the large multisegment ruptures on the San Andreas system;
    2.4. Eliminated the epistemic magnitude uncertainty, which is accounted for by implementing the two magnitude-area relations.
3. Implemented four new recurrence models for Southern California Type-A faults from WGCEP based on moment-balanced models, paleoseismic recurrence models, Ellsworth Type-B (Ellsworth, 2003) magnitude-area relations, and Hanks and Bakun (2002) magnitude-area relations.
4. Developed new multisegment ruptures for several California Type-B faults.
5. Implemented new SCEC CFM model for geometry in Southern California.
6. Revised slip rates for sections of the San Andreas fault, San Jacinto fault, and nine Type-B faults.
7. Developed new zones of distributed shear in Southern California and revised geometry in Northern California.

Fig. 5.2 Significant change in the National Seismic Hazard Map (Credit: US Geological Survey Department of the Interior/USGS/U.S. Geological Survey in Petersen et al., 2011)

with what he has defined as "progressive vision." Meanwhile Coen (2013) has focused on their empirical practices and their definitions of seismic science. Earthquake Junkies fit into all of these categories. As we shall see, they are officially designated scientists and non-scientists alike, advocates, progressives working in a specialized field of hybrid earthquake science.

Indeed, the collaboration of earthquake experts and scientists and their obsession to improve knowledge and policies goes far beyond the walls of any particular ivory tower. As we've mentioned, earthquake studies are produced by many groups and organizations: "state geological surveys, university researchers and research consortia, state and local government agencies, and non-profit and other organizations of the public sector" (Filson et al., 2003).[15] Earthquake experts also work in interdisciplinary fields: they comprise seismologists and earth scientists, social scientists, structural engineers, architects, urban planners, and also policy analysts, lawmakers, buildings contractors, and first responders. They work together to produce the "best available science" (Petersen et al., 2011) of earthquake analysis and natural disaster relief.

[15] The earthquake models have been developed by Caltech scientists in collaboration with UC Berkeley and Stanford University. In addition, the Seismological Laboratory at UC Berkeley has developed close research collaborations with universities in Europe (e.g., ETHZ, Switzerland). State agencies include: the California Earthquake Authority (CEA), the California Emergency Management Agency, Shakeout – The drill, the Earthquake Country Alliance, the National Science Foundation, and the National Earthquake Hazard Reduction Program. In addition, professional associations also play important roles in both research and outreach. For example, the Earthquake Engineering Research Institute (EERI) gathers most of the structural engineers interested in earthquake safety and uses its publication *Earthquake Spectra* to reach out to other experts; the Association of Bay Area Government (ABAG), SPUR, and the San Francisco Planning and Urban Research Association, and private research organizations such as the URS Corporation, an engineering firm, or the RMS Risk Modelization Company have, over the years, recruited some earthquake experts trained at UC Berkeley or Stanford. For decades, the private sector has also been involved in reflections about risk prevention (specifically, large corporations such as Pacific Gas and Electric, Chevron, Google, and others), and public utilities have also been a part of the discussion. Other organizations involved in these discussions include the East Bay Municipal and Utilities District (EBMUD), the National League of Cities, the U.S. Conference of Mayors, and the National Association of Counties, amongst many others.

5.1.3.1 The Organization

Interdisciplinary working groups are the ever-moving forms of organization producing earthquake data for broader communities: during the last decades, each working group (WG88, WG90, WG99, WK02, WK08) has gathered together about 100 scientists to update existing knowledge. (Working Group on California Earthquake Probabilities, 1999, 2003).

Figure 5.3 shows, individual Californian scientists, engineers, and policy-makers, coming from a wide number of academic institutions, private sector government agencies, together with the WGCEP, the CGS, and the SCEC, work to determine the most accurate methodology for developing an earthquake forecasting model. Together, they contribute to the creation of the establishment of the USGS National Seismic Hazard Map, which continues to be updated.

The data production completed by these working groups relies on public funding that has benefited from grants and cooperative financial agreements.[16] Following the San Fernando earthquake in 1971, the US Congress voted to create the NEHRP, which pulled together four federal

[16]The National Science Foundation (NSF) is a fundamental player in the development of earthquake research for the USA, announcing a budget of more than $30 million dedicated to risk and natural hazard research. In 1986, the NSF awarded a five-year grant to the State University of New York, Buffalo, to establish the National Center for Earthquake Engineering Research (NCEER). This grant was renewed in 1991. Another five-year grant was awarded to establish the Southern California Earthquake Center at the University of Southern California. In 1997, the NSF funded three earthquake centers to conduct joint research: the Pacific Earthquake Engineering Center (PEER) sits at the University of California, Berkeley, the University of Illinois-Urbana, and the State University of New York in Buffalo. Acknowledging the challenges of earthquakes in urban area, the NSF has invested $2 million per year for five years in each of the three centers (http://www.nsf.gov/news/news_summ.jsp?cntn_id=102833). New research continues to be important today. After the Haitian, Chilean, and Japanese earthquakes, the NSF awarded three Rapid Response Research grants to gather more information about these earthquakes. Most of this research was created by a consortium of institutions, including, for instance, Ohio State University, the California Institute of Technology, the University of Hawaii, the University of Memphis, and UNAVCO Inc. in Boulder, Colorado. The NSF also supports other programs such as the Incorporated Research Institutions for Seismology (IRIS), which also participates in important research efforts (http://www.nsf.gov/news/news_summ.jsp?cntn_id=116870). One of the most important funding providers is the George E. Brown, Jr. Network for Earthquake Engineering Simulation (NEES), which focuses on earthquake and tsunami loss reduction as it pertains to the US civil infrastructure. Additionally, the NSF has also had a major role in the founding of the National Earthquake Reduction Program.

working agencies to deal with issues and definitions of earthquake risk: FEMA, the National Institute of Standards and Technology (NIST), the National Science Foundation (NSF) and the USGS.

In California, the USGS and its local branch, the CGS; FEMA and its local branch, Cal-EMA; the Seismological Laboratory at UC Berkeley; and the Lawrence Livermore National Laboratory (LLNL) were amongst the first to produce fact sheets and details of earthquake probabilities. Figure 5.3 introduces the agencies present in 2008 and the process of data validation of the NEHPRD program.[17]

5.1.3.2 Hybrid Science and Hybrid Experts

If scenarios are the necessary tool of earthquake risk definition, they are also the result of scientific practices and standards developed in their own times. Such a complex integration of data does not work well without individuals who can establish and maintain the connections between data production and its applicable use, as well as follow the processes of scientific validation.

For some experts, the schematic organization of groups presented in Fig. 5.3 would certainly not work without the energy and dedication of many concerned and motivated professionals, active members in many of these working groups:

> It doesn't work, because if it did, you wouldn't need people like me. All these advocates, the only reason we exist it is because it doesn't work. There are some broken connections or some that never existed. [D.21]

During my many interviews it became more apparent that, in these committees and working groups, science is not only a result of the correlation of objective data collected by instruments, but also of long discussions, arguments, and, finally, consensus.

> Not everything in public policy has science behind it but at least it has consensus. Even if you haven't reduced that epistemic uncertainty, you've got everyone agreeing: it's ok. That's fine, until the next earthquake when they don't feel comfortable anymore. In a way, it's important to have consensus if you don't have science. [D.21]

[17]To reiterate, the NSF is the National Science Foundation, and the NIST is the National Institute of Standards and Technology.

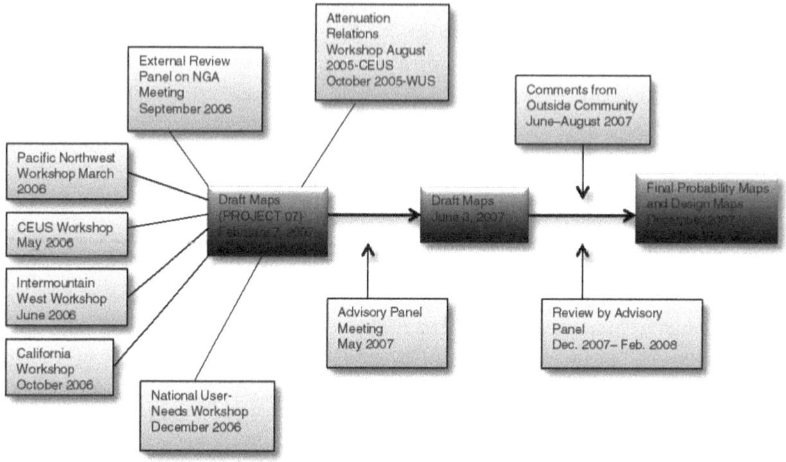

Fig. 5.3 Process for developing the 2008 USGS National Seismic Hazard Map. CEUS, Central United States; WUS, Western United States (Credit: US Geological Survey Department of the Interior/USGS/U.S. Geological Survey)

Committees and working groups also have to deal with changing factors, integrations of new techniques, and new discoveries. No change is small, and everything, from the size of the nails in a building to the quality of the material used to construct beams, can transform—for better or worse—the reaction of a building to a significant earthquake.

> We don't have good wood anymore. It's mostly green and farmed. It's as cheap as possible and it's not the same quality [as] when we cut down old forests. Only the people who observed the wood 30, 40 years ago would recognize that difference. I wouldn't recognize it. [D.21]

This respondent, who works as a structural engineer and is very active in creating public policy blueprints, explained that building public policy on consensus is imperative when conditions of validation cannot be tested. The problem is one of public works and residential buildings: in these cases, consensus is founded on theoretical hypotheses:

> We were talking about the big towers near the [Bay] Bridge, have they been tested? How do you test a 60-story concrete wall? That's why

> buildings are different from automobiles and computers and chairs: you can't build a prototype. It's very important to bring the same kind of thinking to it, but eventually recognize the difference between testing something that has a prototype and something where you have to take a chance. The real world is not our laboratory and we cannot experiment with real projects, that's why we have to be extremely safe. [D.21]

In a field where innovation has been the key in the last few decades, it's impact is difficult to evaluate—that is, until the next earthquake: "If someone has a new idea, where's the experience with it?" [D.21]. Seismology and structural engineering also have to deal with aspects as yet to be revealed—what some experts describe as principles of "uncertainty."

What is discussed in the working groups and committees is a combination of computer generated data and empirical knowledge. The experts' science is a blend of theoretical scientifically verified knowledge together with public consensus, hypotheses, abstractions, and direct observations. The following was heard by a respondent:

> We learn from past disasters, there are also resources such as lost estimations for future disaster scenarios that can be generated by modeling. There is expert opinion we often rely on. If we're not sure how risky a situation is, we can create a committee of experts. We have world-renowned experts who volunteer their services. We go see for ourselves and travel around the world looking at disasters, not just earthquakes. Disasters that would be relevant to California's situation, like Hurricane Katrina. There are a lot of parallels even though the triggering mechanism is different. Recovery issues, emergency shelter issues, market share losses. We take classes; we read and use the internet. We go to conferences. We have mentors and students. [M.3]

Without a doubt, the creation of a relatively new science is a long process, and pertinent knowledge can come from many different sources, materializing in numerous forms. For experts and scientists working on earthquake definitions, the world becomes a de facto laboratory, one in which their experiences and understandings of ramifications regarding risk prevention and disaster mitigation is determinant.

To answer the question "How did you get your knowledge of risk and disasters?," a longtime expert who participated in the writing of several reports answered clearly: "through experience" [T.4]; another expert added that there is "no substitute for seeing actual damage" [D.21]. The earthquake community in the Bay Area relies on the hands-on competencies and true-life experiences of its members to make this schematic organization function efficiently and usefully. Asked how he navigates and collects relevant information within this complex network of organizations, commissions, working groups, and so on, another respondent involved in drafting legal recommendations answered with a laugh:

> People I rely in people! I will use the network to know where people work and I listen to their experience. I also scan the literature [...] and I participate in organizations like FEMA and others that give information. This is the kind of work that I do, I look for information and I comply. [T.4]

Because much field research has been conducted following the "snow ball method,"[18] where one interview leads to another and makes it possible to follow the actants, I noticed that individuals with long-term activity in the seismic field were often referred to as important resources for the entire community of earthquake scientists and experts. This observation can be heard from people who refer me to additional experts for information on a specific topic: "He is the wisest of all," [J.14] said one respondent of another authority. Indeed, legitimacy inside this extended community has been an important aspect of cohesion and respect amongst many professionals working in the field: "You know she's been in this field a long time. [...] So, she's very good." [R.5], one interviewee said to me at one point.

The recognition of a community of experts is part of a broader acknowledgement of the ability for many to work together and to solve the multiple issues that earthquakes pose to Bay Area residents. "He's totally into it, he is devoted, you know, and he spends a ton of time on this, it's a passion of his, it's really great" [J.14], one respondent remarked. Within this context, experts themselves redefine their definitions of expertise, recognizing the multi-dimensional aspects of knowledge needed to build a fuller understanding of earthquake risk.

In so doing, they put into circulation the division between experts and amateurs, which is a trend that was developed at the very beginning of Californian seismology. During my field research, people would often

[18] See Introduction regarding the methodology.

refer to a specific example of this fluidity; this can be heard, for example, in the following story told by a USGS scientist:

> The city of Berkeley 20 years ago there was a mother whose daughter was in school and she asked the school how safe the building is and they couldn't answer, at the time I'm not even sure she was employed, she's basically a mom, she caused most of the buildings to be retrofitted and some to be condemned and she ended up working in the city offices as assistant state manager after all this. She's still very active but she's at Harvard, she got into it just because she asked a question about her daughter's school, now 25, 30 years later she's one of the experts. She was very involved in getting Berkeley city hall retrofitted and fire stations and schools, it really does take individuals to go and make things happen. [K.11]

Scenarios are the visible parts of earthquake experts' and scientists' "best practices." They are the end results of many different pieces of knowledge that have been discussed, vetted, and balanced to the point that they can be considered strong enough to be good actants in the context of risk prevention and seismological knowledge. During that long process, categories of knowledge and subjects that were once thought immutable are transformed to help produce the best science possible.

5.2 A Step Further: Operations of Translation

Building on the necessity of this hybridization of science and experience, it is necessary to focus on three other operations of "translation" which address questions of territories of experience, general practices, and knowledge. Souriau noted, "Generally, one can say that to know what a being is, you have to instaure it, even construct it, either directly (happy are those, in this respect, who make things!) or indirectly through representation—up to the point where, lifted to the highest point of its real presence and entirely determined by what it thus becomes, it is manifested in its entire accomplishment, in its own truth" (Souriau, 2009). As Michel Callon writes, "Translation is a process, never a completed accomplishment, and it may [...] fail" (Callon, 1986). In this section, I focus more precisely on three operations that allow for the emergence of earthquake risk analysis in order to see how different forms of data collection and interpretation have shaped our understanding of the risks of earthquakes and the various methods established to mitigate their negative effects.

5.2.1 *Translation 1: From Event to Knowledge*

As one seismologist at UC Berkeley stated, "in seismology, discoveries are driven by events, by big earthquakes" [R.9]. As discussed, the work of translating earthquakes into scientific experience is a long process.

In fact, it was only after tectonic plate theory[19] was accepted in the 1960s that seismologists were able—and still, today, are only partially able—to describe and explain the mechanisms that trigger an earthquake. Some of the earliest known scientific comments regarding earthquakes occurred in the mid-1600s. Most historiography on seismology starts at after the Lisbon earthquake in 1755. In the early 1800s, the first observations conducted by seismologists helped to define what today are still called the P and S waves (pressure and secondary waves). In 1857, the Naples earthquake gave scientists the opportunity to make attempts at observational seismology. Following this, the first seismographs were built by British scientists in Italy and in Japan. In 1897, North America's first seismograph was installed near San Jose, California; this later recorded the 1906 San Francisco earthquake. The history of seismology, starting at the beginning of the twentieth century in California, has been a long history of instauration of a clear—or, given various limitations, the clearest possible—science.

In California, the first Big One was an 1868 earthquake,[20] which held the title for decades before the 1906 earthquake hit. In the still very rural Bay Area, this earthquake produced little damage and was left relatively undocumented. However, the first seismometers were installed in the Student Observatory on the UC Berkeley campus a decade later, in 1887. Discussing this history, one seismologist noted:

[19]The theory of tectonic plates between 1956 and 1967 (Frankel, 2012) validated that earthquakes are the results of movements created by the accumulation of energy in the always moving Earth crust.

[20]Most of the documentation about the 1868 earthquake was produced in the twentieth and twenty-first centuries.

> The first seismic networks were global seismic networks, there would be an observatory, and the first seismic observatory in the western hemisphere was here in Berkeley. That was in 1887 or something like that, the way that seismology started was you have a few seismic observatories around the world and between them you can detect big earthquakes. [R.9]

After the 1906 earthquake, the Lawson Commission's report was the first full-scale attempt at the comprehensive documentation of an earthquake: "It afforded an exceptional opportunity for adding to our knowledge of earthquakes" (in Lewis, 2008), noted geologist Andrew Lawson, Head of the Commission. Lawson, who was Chair of the Department of Geology at UC Berkeley at the time, dispatched teams on foot and horseback to explore the fault, from Humboldt County in Northern California all the way to Coachella Valley, south of Los Angeles. By 1908, Lawson had graphed almost the entire San Andreas Fault while looking at the 290-mile rift caused by the 1906 earthquake. He then produced a careful, detailed report that included the elastic–rebound theory, which was an important step in instauration of the earthquake as an object of science:

> According to these theories, earthquakes were due to the sudden release of strain that had been gradually built up by the constant creeping of the earth's surface near a fault. In his contribution to the commission's final report, Harry Fielding Reid had argued that there had indeed been a gradual distortion of the earth's surface near the San Andreas Fault during the late nineteen century, just as the elastic rebound theory called for. (Geschwind, 2001: 60–61)

Prior to the publication of the Richter scale in 1935, and the increasing instrumentation of known, active faults, or even the more recent computerized modeling techniques, seismologists have relied on many eclectic methods of analysis. At its nascence, when science relied mainly on the description of trained observers, no detail was too small; no nuance in the experience of a felt earthquake was considered too trivial in the construction of this new science. Operations of translation from an observed event to "scientific" knowledge prevailed in the recognition and definition of earthquake risk.

> By the early twentieth century, the availability of seismographs allowed many earth scientists to dream of turning their discipline into a quantitative, objective science, modeled by physics. They transformed what counted as evidence of the earth's history. Out went data filtered by human bodies; in came the hard evidence of seismographs and accelerometers. (Coen, 2013: 20)

The human body was the first agent of translation, before the use of machines. Whether interpreting their own observations or relying on the traces of a seismograph, seismologists made connections in order to establish relations between experience and science. They learned how to "read" an earthquake (November, Penelas, & Viot, 2009) to give sense to feelings and traces regarding a particular seismic phenomenon.

In 1931, Harry O. Wood, who had been working for decades with eyewitness earthquake observation reports, also known as "felt reports", published the "Modified Mercalli Scales" with Franck Neumann. This new scale was designed to make reporting easier, to define the earthquake with degrees and thresholds, eliminating ambiguities, but also to "insert explicit statement about the mental states conducive to certain reported effects" (Coen, 2013: 258) (Fig. 5.4).[21]

The work of translation was important, and Wood struggled for a decade to find the appropriate vocabulary with which to describe the variously seen, felt, and heard effects of earthquakes:

> Though I have worked hard on this scale, it is still containing inconsistencies [...]. I have never experienced an earthquake in which all work of construction was greatly damaged or destroyed, but such earthquake has occurred. Such fault slips as occurred in 1906 or in 1915 must be indicative of great energy whether or not productive of great intensity. There were anomalies in 1906 and in any experience there always are.... No scale is or can be perfect, but before we adopt this one or accept any other change we must do our best to adjust the definition and eliminate any absurdity, bearing in mind that some anomaly will always be found. (Wood as cited in Coen, 2013: 260)

Based on the scale described above, Wood developed a large cooperative network of informants, the "citizen-observer," as Coen

[21] The Modified Mercalli Intensity Scale is still used today in the Shake Map, also known as the "Did You Feel It?" map.

Example of modifications to the Californian section of the Map:[1]

1. Revised earthquake catalog and accounted for magnitude round off and uncertainty.
2. Constrained model to fit within two sigmas historical or observed seismicity rates (suggested by WGCEP—Science Review Panel):
   2.1. Reduced moment rate on faults by 10 percent to account for aftershocks, foreshocks, after slip, and smaller earthquakes;
   2.2. Reduced earthquakes M6.5 in smoothed gridded seismicity to one-third the rate to account for earthquakes already modeled on faults (generally not applied outside California);
   2.3. Implemented a branch of Gutenberg-Richter Model with b=0, which is consistent with modeling of several of the large multisegment ruptures on the San Andreas system;
   2.4. Eliminated the epistemic magnitude uncertainty, which is accounted for by implementing the two magnitude-area relations.
3. Implemented four new recurrence models for Southern California Type-A faults from WGCEP based on moment-balanced models, paleoseismic recurrence models, Ellsworth Type-B (Ellsworth, 2003) magnitude-area relations, and Hanks and Bakun (2002) magnitude-area relations.
4. Developed new multisegment ruptures for several California Type-B faults.
5. Implemented new SCEC CFM model for geometry in Southern California.
6. Revised slip rates for sections of the San Andreas fault, San Jacinto fault, and nine Type-B faults.
7. Developed new zones of distributed shear in Southern California and revised geometry in Northern California.

Fig. 5.4 Abbreviated Modified Mercalli Intensity Scale—with correlation in Magnitudes (Credit: US Geological Survey Department of the Interior/USGS/U.S. Geological Survey)

(Coen, 2013: 256) describes them, many of whom were women. Working hand-in-hand with reporters, Wood was also very impressed by the quality of the observations that newspapers published: “I find that several of the newspapermen have in Imperial Valley had printed very good accounts of the recent shocks and they appeared to be very interested in the whole subject” (Coen, 2013: 257). During his active years, Wood maintained a large correspondence with observers, graciously answering their interrogations, adding to their comments, and, more importantly, encouraging them to pursue their own observations.

These efforts paid off after the Santa Monica earthquake struck in 1930, when more than 300 reports were sent to the researchers from “private individuals, postmasters and other public officials, and representative corporation cooperating with the Survey” (Wood as cited by Coen, 2013: 256). Nonetheless, experts with little experience dealing with public participation started debating the validity of this method of data accumulation, as well as questioning the scientific basis of the results.[22]

[22]As Coen states, the controversy follows the line of a Kantian definition of science (Coen, 2013: 254).

At the time of the publication of this scale, seismography was very much considered an imperfect science. Wood and Neumann noted that, "though the importance of the factor of acceleration is recognized, we have as yet no satisfactory definition of intensity, no formula expressing earthquake violence in terms of ground movement" (Wood & Neumann as cited in Coen, 2013: 259). For this reason, Wood encouraged the young Charles Richter to focus on this particular problem: namely, creating a mechanical equivalent of intensity. In 1928—just as occurred nearly a century later, during the 2011 Fukushima earthquake—plate movements were found to be too strong to be registered by seismographs.

Several years after the Long Beach earthquake of 1933, Richter started to think about a scale of measurement that would help reduce misinterpretations of the size of an earthquake, as well as to make better and easier comparisons between earthquakes. Coen noted the Richter Scale—which had originally been created for local, relative, and subjective uses by earthquake eyewitnesses—had later become a universal and absolute measure of the violence of all earthquakes. The 1933 Long Beach earthquake was also a wake-up call for the seismology community, who realized that this particular earthquake had been far more destructive than it should have been; "It was clear to Wood that foolish building practices were to blame" (Coen, 2013: 261). Wood, used the term "apparent intensity" (Coen, 2013: 261) to describe the fact that the amount of damages from this earthquake were not necessary correlated with the force of the earthquake.

In the first part of the twentieth century, Wood and Richter were responsible for providing an interpretative framework for earthquakes; in other words, they allowed for a way to translate a particular earthquake into words and then place it on an intensity scale. A century later, this interpretative move is still a work in progress, as contemporary seismologists, using new technologies, continue to revisit past events and engage again with both scientists, experts and with the public. During one of our meetings, a seismologist at the Seismological Laboratory at UC Berkeley recalled:

> The 100th anniversary of the 1906 Great San Francisco Earthquake and the 140th anniversary, in 2008, of the 1868 earthquake on the Hayward Fault have been the occasion for experts and non-experts to meet and both develop new models and improve reach-out activity toward the

population. These commemorations have been also[23] to develop 39 scenarios involving different earthquake features (such as variable magnitude, distribution of slip, location of the epicenter and speed of propagation) and their consequences in the Bay Area. Along with giving more fuel and data to work with, these scenarios have also reinforced feelings of uncertainty of the future and the necessity to pursue the work and reach out to the research community. [P.1]

Despite the tremendous progress realized in the twentieth century regarding the understanding of earthquakes, much remains to be done. Instrumentation has overtaken eyewitness felt reports but, despite this considerable progress, earthquakes remain hard to grasp. As one seismologist confessed:

> We think we understand where all the faults are so we know where they're going to occur, but both the Northridge and Loma Prieta earthquakes occurred on unknown faults. That was a surprise to me professionally. [R.9]

5.2.2 Translation 2: Monitoring the Earth's Crust

In the last few decades, the development of seismology has brought together a number of disciplines that have joined forces to better understand the unfolding nature of earthquakes. In 1998, the NSF established the collaborative Network for Earthquake Engineering Simulation (NEES) with 14 research centers that share a centralized data repository and earthquake simulation software.[24] To guarantee progress, this research consortium has heavily relied on the networks of seismographs and GPS devices, which have been used continuously since the 1960s. As one researcher explained:

[23] USGS, URS Corporation, Lawrence Livermore National Laboratory (LLNL), Stanford University, and UC Berkeley.

[24] Cornell University; Lehigh University; Oregon State University; Rensselaer Polytechnic Institute; SUNY, Buffalo; University of California, Berkeley; University of California, Davis; University of California, Los Angeles; University of California, San Diego; University of California, Santa Barbara; University of Illinois, Urbana-Champaign; University of Minnesota; University of Nevada, Reno; and the University of Texas, Austin.

> What is called the modern seismic era is based on regional networks that started in the 1960s, 1970s. That's what detects earthquakes; that's what allows us to understand the distribution and recurrence interval of earthquakes. [R.9]

Today, these programs gather an impressive amount of data—shared across laboratories and universities—on a number of aspects of plate movements. Most of the data are collected through instruments that have been installed across California in the last few decades. This network is largely the result of advocacy actions taken by earthquake researchers after some of the last major California earthquakes:

> The infrastructure we use for seismology and physical networks have significantly improved after the last big earthquakes. The network we have today came in following the 1989 earthquake. In Southern California, they have a much bigger network because their most recent earthquake was in 1994. [R.9]

Although most of the instruments that help define and forecast earthquakes in California have been installed along California faults, the research network that uses this data overlaps their large geographical boundaries. Further, most of the data collected in California has been made available to researchers worldwide. Today, data used by earthquake scientists come from different systems of measurements: as we shall see, each system allows us to look at the faults from different angles and perspectives—again, utilizing different scales.

5.2.2.1 Tool Box

As other types of science progress, seismology benefits from developments made in other domains. The introduction of more sophisticated computers has deeply transformed the nature of this research, relying more on today's algorithms than on yesterday's field trips made on horseback. As one expert explained, being able to collect data and translate it for so many actors has been a turning point in understanding the mechanics of a given earthquake:

> Understanding the [earthquake's] process is not being purely random, but there's this kind of machine behind them. From that point, we do a good job of understanding how the energy is radiated during an earthquake

> and being able to both predict earthquake scenarios and the distributional ground shaking. For earthquakes that have just happened, we are also able to collect data, combine them with physical models and map them to compare what we thought the ground shaking was. [R.9]

The data provided by this network is utilized by interdisciplinary programs, such as those developed by the Pacific Earthquake Engineering Center (PEER). PEER illustrates the diversity of competencies engaged in earthquake research: geotechnical modeling, performance-based seismic design methods and tools, testing and characterization of structural foundations and systems, soil mechanics and bituminous materials.

In northern California, collecting and treating seismological and geodetic[25] data is an activity shared by several institutions, including UC Berkeley's Seismological Laboratory and the USGS, grouped together under an umbrella organization called the Northern California Earthquake Data Center (NCEDC).[26] Data is collected through 11 networks that produce other types of information: maps of strong ground shaking, "seismo-cams," earthquake reports, waveform data, and so forth. All of this information is grouped into a catalog of locations, magnitudes and phases, and moment tensors,[27] as well as the Townley-Allen Catalog, which is based on newspaper accounts covering the long period between 1769 and 1927.

Other, more local networks, such as the Bay Area Regional Deformation Network (BARD), started in 1991, is a network of 32 continuously operating GPS receivers at various sites in the Bay Area and northern California that measure slip occurring on various fault lines. As it states on their official website: "The primary goal of the network is to monitor crustal deformation across the Pacific-North America plate boundary and in the San Francisco Bay Area for earthquake hazard reduction studies and rapid earthquake emergency response assessment".

The Berkeley Digital Seismic Network (BDSN) also includes stations that record data for the Mini-Plate Boundary Observatory Project, which looks at plate boundary deformations. Here, again, the presence of heavy instrumentation is essential for the success of earthquake measurement. As one seismologist shared:

[25] Geodetic study measurement and representation of the Earth, such as crustal motion.

[26] The Southern California Earthquake Data Survey covers the southern part of California.

[27] Mathematical representation of the movement of faults during an earthquake.

> When I say the network, I mean the physical seismic network. Berkeley runs about 50 seismic stations across northern California. There are all these seismic stations throughout the region and they all stream their data back to here [the Seismological Laboratory at Berkeley]. That's just a physical infrastructure, it's like a telescope, but it's just collecting data, so some of the funding is just to run that network to collect that data. [R.9]

To get even more measures, those coming from deep underground, the Plate Boundary Observatory Network (PBON) has been installing borehole strain meters to measure very small changes at depths ranging from 100 meters (328 feet) to 250 meters (820 feet), as well as continuous GPS measurements around the Bay Area.

Earthquake researchers also use satellite technology: Light Detection and Ranging (LiDAR) and Interferometric Aperture Radar (InSAR) provide information to refine earthquake probability estimations. LiDAR,[28] which is used in aircraft and combined with GPS technology, is a system that allows for the detection of a topographical system. In 2012, LiDAR revealed previously unknown faults near Lake Tahoe.[29] The InSAR system is satellite-based radar that evaluates ground surface movement and its transformations. The combined data have allowed seismologists to distinguish different fault motion speeds across the Bay Area. Especially adapted to take measurements in urban areas and in challenging conditions (e.g., bad weather or at night), they have opened up new ways to look at seismic movement.

Seismic movement—or "creep"—is especially important with the Hayward Fault, which, unlike the San Andreas Fault, releases stress through slow and steady motion that causes deformation in street curbs and buildings.

Finally, looking at building reactions to earthquakes, the Advanced National Seismic System (ANSS) has placed seismic monitors on several thousand buildings. These monitors report information such as soil conditions, distance from ground ruptures, and the length and intensity of shaking from tremors. These devices, designated to provide extra

[28] Also called LADAR (Laser Detection and Ranging in military contexts).

[29] LADAR Technology Reveals faults Near Lake Tahoe. Released: 5/23/2012 2:33:42 PM, U.S. Department of the Interior, U.S. Geological Survey, Office of Communications and Publishing (http://www.usgs.gov/newsroom/article.asp?ID=3218#.T_9bBXCiqjw).

information on effects such as the high amplification of seismic waves, also help to improve future designs of seismically sound buildings.

5.2.2.2 Failures of Translation: The Parkfield Experiment

Being able to translate the Earth's movement into data is not an easy operation and heavy instrumentation in the field does not always help: in fact, in California, most of the recent earthquakes have happened along previously unknown fault lines—the Coalinga earthquake in 1983, the Whittier earthquake in 1987, and the Northridge earthquake in 1994, as well as small earthquakes that went undetected by the technology of their times. Nonetheless, most earthquake scientists remain confident that this enormous body of data will help them define, and even improve, earthquake forecasting.

However, the experience drawn from the Parkfield Experiment attests that this extensive data collection does not always bring the anticipated results. Being forced to give up the dilatancy-diffusion hypothesis of the 1970s,[30] earthquake researchers have now returned to empirical research—but, this time, with as many instruments as possible.

Geologists discovered that, close to the town of Parkfield, California, on a segment of the San Andreas Fault midway between Los Angeles and San Francisco, M5.5 earthquakes struck at regular intervals in the years 1857, 1881, 1922, 1934, and 1966. Because the 1934 earthquake was believed to have happened a decade "in advance," researchers thought they had found a place where recurring patterns could, if not proven, at least be tested. The wave forms recorded in 1922, 1934, and 1966 suggest that the ruptures happened in the same part of the fault.

Earthquake researchers who were interested in looking at the precursors of earthquakes predicted a 90–95% chance of a M5.5 earthquake occurring between the years 1983 and 1993. Waiting for this earthquake, they installed seismometers, creepmeters, and strain meters along this part of the San Andreas Fault.

[30] "The dilatancy–diffusion hypothesis was one of the first attempts to predict the form of potential geophysical signals that may precede earthquakes, and hence provide a possible physical basis for earthquake prediction. The basic hypothesis has stood up well in the laboratory, where catastrophic failure of intact rocks has been observed to be associated with geophysical signals associated both with dilatancy and pore pressure changes. In contrast, the precursors invoked to determine the predicted earthquake time and event magnitude have not stood up to independent scrutiny" (Main, Bell, Meredith, Geiger, & Touati, 2012).

> As a prediction experiment, the principal goal of the Parkfield study is a detailed description of the final stages of the earthquake preparation process; observations at Parkfield should aid in the evaluation of the feasibility of intermediate—and short-term earthquake prediction elsewhere. Furthermore, the detailed history of strain accumulation and release over a complete cycle that is being recorded at Parkfield should provide the basis for testing and refining models for earthquake recurrence on plate boundaries.

This experiment, which was also designed to be one of the first steps towards an early warning system, was intended to provide information to the Governor's Office of Emergency Services. Unfortunately, the setting failed to produce the desired results. As one historian writes:

> The coming Parkfield earthquake promised to provide valuable data on earthquake precursors. In order to record these data in sufficient detail, earthquake researchers turned the Parkfield fault segment into the most densely instrumented segment in the country. To their great chagrin, however, the expected Parkfield earthquake by the end of 2000 still had not occurred. (Geschwind, 2001: 216)

The USGS, who installed the instrumentation of the fault, was strongly criticized for the inaccuracy of its prediction, which was called "the geophysics' Waterloo." In fact, some went so far as to predict the "defeat" of science[31]. However, since the installation of the first instruments, Parkfield has remained a site of experimentation. In 1994, a working group concluded that, despite the lack of an earthquake at Parkfield, the experiment has not yet failed and should continue to be pursued as "Parkfield remains the best identified locale to trap an earthquake".

In June 2004, the USGS and NSF started another program: the San Andreas Fault Observatory at Depth (SAFOD) program. In September of that same year, the long-delayed M.6 earthquake finally struck near Parkfield, 16 years too late, proving yet again that earthquakes are difficult phenomena to grasp and to "trap."

[31] In an article entitled "Small earthquake somewhere, next year – perhaps," a journalist from *The Economist* wrote: "If the earthquake comes there without warning of any kind, earthquakes are unpredictable and science is defeated. There will be no excuses left, for never has an ambush been more carefully laid for such event".

5.2.2.3 New Developments, Old Patterns

The introduction of new software and new communication technologies in recent years has opened up a new chapter in earthquake technology: the possibility of early warning systems.

> What we're working on right now is the idea of being able to do that [early warning system] before people feel the shaking, so when the earthquake begins you detect the initial low amplitude shaking close to the epicenter and predict the total area that's going to experience the shaking. [R.9]

In September 2013, the Governor of California, Jerry Brown, signed a bill which requires the Governor's Office of Emergency Services to develop an early warning system while the state continues to search for the $80 million needed to finance this operation. Led by the Seismological Laboratory at UC Berkeley, this new development will rely on recent capacities to process and communicate data through contemporary telecommunication means (e.g., cell phones, Internet broadcasts, and other dedicated communications systems for special users, such as airports, trains, utility companies, and so forth). For researchers, this was:

> A fundamental change in seismology. […] What this would do is it would take our products, what the seismology community is doing with this data and it would appear on everybody's cell phone, so the impact of seismology on society would be huge because, all of a sudden, everybody's getting this information, whereas at the moment, the way that society interacts with seismology is passive. Seismologists come up with estimates of ground shaking and then it's the engineers that take the ground shaking and use them in buildings. This would be a direct communication between seismology and the community as a whole so it'd be a big change. [R.9]

At this stage, the project is being developed mostly in collaboration with large corporations and utility companies around the San Francisco Bay Area, including Google, Chevron, PG&E, and BART, amongst others.

Looked at through the lens of history, the early warning system is another major step in the earthquake researchers' ambition to anticipate—predicting or forecasting—seismic events. Simultaneously, it is also another major step in the history of the legislative effort to assess and deal with earthquake risk.

5.2.3 *Translation 3: A Step Towards Safety?*

Defining earthquakes via instruments has often been found to be more complex than expected, and the lack of success good reminders of the difficulties in embracing the complexity of earthquake phenomena. However, despite these earlier disappointments, experts' continuous efforts to understand the mechanisms of earthquakes and their consequences for built environments have created and developed further the regulatory apparatuses designed to strengthen building construction codes in California, and to avoid, as much as possible, the major damage caused by an earthquake.

> The story of earthquake hazard mitigation in twentieth-century California, then, is the story of the Progressive impulse among a small group of California scientists and engineers and the persistence of that impulse even after other reformers had become disenchanted with the state. (Geschwind, 2001)

Published regulations for building construction are most likely those things that have contributed the most to earthquake safety in the Bay Area. But, considering the limited audience that such publications reach, their important impact remains largely unknown by the general public.

5.2.3.1 Creation of a Legislative Body

At the beginning of John Fante's 1939 novel *Ask the Dust*, the narrator, Bandini, describes his anxiety in front of a large concrete building, which were very common in Los Angeles before the Second World War: "These people walking in and out of huge concrete buildings—someone should warn them. It would come again; it had to come again, another earthquake to level the city and destroy it forever. It would happen any minute. It would kill a lot of people, but not me" (Fante, 1939). To avoid such a tragedy, and for nearly a century afterwards, governments—local, state, and federal—have taken important actions to minimize damage, injury, and death.

Earthquake experts have long been advocates of regulatory solutions to minimize risks caused by unsafe buildings. Structural engineers first began to organize after the 1906 San Francisco earthquake. At a time when city officials were trying to erase the marks of the earthquake, engineers and seismologists were trying to slow down the real estate businessmen's appetite for increased space and resources.[32] For these earthquake experts, it was also important to take the actions that were not taken after the 1868 earthquake: study of earthquake phenomena in order to improve building design.

Within a couple of years, seismologists and engineers started to work in tandem to increase the number of safely built buildings in California. To accommodate this objective, Professor Lydik S. Jacobsen invented the first "shaking table" at Stanford—a large platform on a wheel that could be made to vibrate—and the professional organization of structural engineers was founded in order to share resources and knowledge.[33]

The event that gave some visibility to earthquake experts and allowed their wish for safer buildings to be constructed was the M.6.4 Long Beach earthquake on March 10, 1933, at 5:55 p.m. The earthquake struck southern California and, while damages were limited, a total of 15 of 32 Long Beach schools were badly damaged. Newspapers relayed the outrage of parents. Harry Wood, told reporters that:

> The practical lesson of the recent temblor, as of all others, [...] is build well and choose or prepare strong foundations. Design for strength and [construct] conscientiously using good materials. Avoid what experience has shown to be faulty. (Geschwind, 2001: 107)

The event marked the first *modus operandi* between organizations that continues today: the work by joint committees that move progressively from the local level to the state, then up to the federal government. The same year, in 1933, discussions were held which ultimately led to the Field Act[34]

[32]The construction of the city of Chicago inspired the reconstruction of the San Francisco, including the use of steel frames and masonry in-filled walls.

[33]In 1930, the Northern California Association of Structural Engineers was officially launched, with 31 members. Soon, the Northern and Southern Associations joined forces to form the Structural Engineers Association of California (SEAOC).

[34]Assemblyman C. Don Field, a Republican, prepared the Field Act. For Turner, "The Field Act transferred the regulation of public school design and construction from local governments to the State's Division of Architecture. Its enforcement on over 70,000 school construction projects since then has generated significant improvements in the practice of earthquake engineering" (Turner, 2004: 4). As we shall see later, the Field Act was seen to be a protective shield for many Californians regarding the soundness and safety of their children's schools; unfortunately, a large-scale investigation conducted by California Watch revealed that the Field Act was not enforced in most schools in California.

and the Riley Act,[35] two laws that focused on the safety of public buildings, and attempts at trying to avoid the proliferation of unreinforced masonry buildings.[36] Since 1933, more than 80 laws have been passed in California to promote earthquake safety and to mitigate damage and injury.

Addressing what has become a common concern in California, the first seismic code, known as the Building Code of California, was published in 1939,[37] after more than years of work by small committees.[38] The expanding needs of the Bay Area for building regulations in the 1940s[39] was the result of a real estate boom then occurring throughout the state.

Thus, together with seismologists, structural engineers significantly participated in the development of the earthquake field. With the influence of these professionals, lightweight concrete became the preferred material to use with steel-framed buildings; with this, the standardization of structural requirements was on its way. In the following decades, the building codes became progressively stronger in accordance with new discoveries after each earthquake. As one structural engineer recalled:

[35] Approximately 10–15% of California's buildings were constructed before 1933, at a time when few cities had building codes. The Riley Act "requires local governments to have building departments that issue permits for new construction and alterations to existing structures and conduct inspections. Permit fees paid by building owners generally fund the work of local building departments. The Act also set minimum seismic safety requirements that have since been incorporated into all building codes." Western States Seismic Policy Council (http://www.wsspc.org/policy/California.shtml).

[36] "An unreinforced masonry building (or UMB, URM building) is a type of building where load bearing walls, non-load bearing walls or other structures, such as chimneys, are made of brick, cinderblock, tiles, adobe or other masonry material, that is not braced by reinforcing beams. The term is used in earthquake engineering as a classification of certain structures for earthquake safety purposes, and is subject to minor variation from place to place. URM structures are vulnerable to collapse in an earthquake. One problem is that most mortar used to hold bricks together is not strong enough. Additionally, masonry elements may 'peel' from the building, and fall onto occupants or passersby outside" (Wikipedia Contributors, 2018b).

[37] A first edition of the Uniform Building Code was published in 1927. The eastern part of the United States has had their building codes in place since 1914.

[38] It is interesting to note that committees remain the principal form of organization for structural engineers and the seismic community in general. Most of the reports that have been produced in recent years have been done so via committee consensus.

[39] The Vesano Building Code, named after Harry Vesano, former Director of Public Works in San Francisco, was published in 1948.

> Some engineer realized that if you have a frame where you have two columns and a beam, the beams have to be enormous and the columns are enormous. When you get that big, the welds between the beams and the columns don't work. When you optimized, you're creating a condition where it can't work in [earthquake] conditions. After the earthquake, engineers noticed that [damages] might have to do with that. All our code provisions were written on the presumption that beams and columns were everywhere, but they didn't speak up in advance. [D.21]

Until the 1970s, local authorities published their own building codes, drawn from recommendations by UC Berkeley specialists, together with local amendments. The California Building Standards Code (California Code of Regulations, Title 24) was created in 1978 as an amalgamation and reorganization of existing codes. In 1986, the State of California passed the "Unreinforced Masonry Building Law" (SB 547), which required each jurisdiction in California's "Seismic Hazard Zones" to reduce unreinforced masonry hazards, which remains today one of the most hazardous types of building in the event of earthquakes. Over the last few decades, California's buildings codes have continued to be revised several times, and are currently published on a triennial basis.[40]

Part science and part experience, the buildings codes today are subject to criticism and discussion that takes place in private and semi-private settings with a large number of committees:

> The creation of the building code is generally consensus processes. It comes out of committee, and it has good things and bad things. Policy in the end is how good a building should be, has not been resolved; it's being debated today. The answer is it's as good as it used to be. We avoid dealing with that policy in the building code. We don't know what fine means but you'll be fine, the purpose of the building code is to comply so you're legal and can't be sued. If you want to do something outside these rules, you can do that but you have to prove its equivalent. If you build across the bridge you'll see some of those tall buildings, the engineer came a few years ago and said: "we want to use a new system" [...]. The

[40]As one expert recalled: "Something else that's been done in the United States is that we used to have three building codes, we used to have one for the Northeast, for the South, and one for the West. And after Loma Prieta, FEMA said, 'We can't deal with three codes, we go around the country trying to assess buildings and there are different codes so, uh, we now have the international building code and it is the code that's used throughout the country. So that's changed." [T.4]

> engineer comes and says we're going to do higher than 240 feet and its equivalent and it started a discussion in city council about the code and how to benchmark it. That's a question that's never been solved. [D.21]

For researchers, the buildings codes have had a considerable impact on the evolution of the built environment. However, research have shown that the building codes, as good as they could be, were often not enforced by local building departments (Burby & May, 1998). For the structural engineer Fred Turner, "The greatest challenges to California's building departments continue to be: 1) The lack of public awareness and political support for effective code enforcement; 2) The risks of existing vulnerable buildings in earthquakes; and 3) The need to enhance staff sizes and qualifications to keep up with phenomenal growth" (Turner, 2004: 10). Turner, now in charge of a state organization dedicated to earthquake prevention, explained that the process of making law is a very long one:

> [After] the 1971 San Fernando earthquake, the State of California realized that hospitals are facilities that really should be standing up after an earthquake. And they transferred the design of hospitals to an agency called OSHPD – Office of State Planning and Health [sic].[41] Every hospital has to go through a plan-check and a design-check by OSHPD. New hospitals are doing pretty well, but we still have this huge inventory of hospitals that were designed before [1971]. Finally, 1989 happens, the Loma Prieta earthquake, and once again, people are thinking, "okay, our hospitals are still vulnerable...," and so they started to—this is just the State of California—create a group that looked into a Senate bill. They eventually passed SB1953,[42] and came up with deadlines for when you need

[41] Office of Statewide Health Planning and Development.

[42] SB1953 was, in fact, written in 1994 after the Northridge earthquake as the continuation of the Alfred E. Alquist Hospital Seismic Safety Act, also known as the "Hospital Act". "The Alfred E. Alquist Hospital Seismic Safety Act ("Hospital Act") was enacted in 1973 in response to the moderate Magnitude 6.6 Sylmar Earthquake in 1971 when four major hospital campuses were severely damaged and evacuated. Two hospital buildings collapsed killing forty-seven people. Three others were killed in another hospital that nearly collapsed. In approving the Act, the Legislature noted that: '[H]ospitals, that house patients who have less than the capacity of normally healthy persons to protect themselves, and that must be reasonably capable of providing services to the public after a disaster, shall be designed and constructed to resist, insofar as practical, the forces generated by earthquakes, gravity and winds.' (Health and Safety Code Section 129680) When the Hospital Act was passed in 1973, the State anticipated that, based on the regular and timely replacement of

to evaluate your hospital, when you need to have it "life-safe," and when you need to have it meet full code. And this was 20-some-odd years after the 1971 earthquake. So, if you look at [it], from start-to-finish, that's a 50-year process, almost. So, it's huge. But that came from the top down. And the reason that it takes so long is that it's outrageously expensive, it's billions of dollars. [J.15]

Structural engineers and seismologists are very active in the adoption of state laws that require California to build more earthquake-resistant buildings, specifically through housing policy, good governance, and infrastructure policy (ABAG, 2013a). But these processes often span beyond the career of any single individual and need to be carried over by one generation of passionate advocates to another.

5.2.3.2 Cities' Resilience

Since the time of the last few major earthquakes, Earthquake Junkies have been working on setting standards for municipalities, looking at local past disasters, but also with a broader view of other significant events that have transformed our understanding of many different types of risk and disaster: the Kobe earthquake; the events of 9/11; Hurricane Katrina, and, recently, the Haiti, Christchurch, and Fukushima disasters.

As working groups are at the heart of defining earthquake risk for scientists, the committees have become flexible enough to consistently translate potential hazards into possible real-world risks. Two specific organizations involved in the definition of resilience in the San Francisco Bay Area are the Association of Bay Area Governments (ABAG), an advisory organization and regional planning agency that was originally formed in 1961, bringing together representatives from the nine counties that surround the Bay Area, and SPUR, previously known as the San Francisco Planning and Urban Research Association, a non-profit research education and advocacy organization, which was originally formed after the 1906 San Francisco Earthquake by city leaders looking to improve housing quality.

aging hospital facilities, the majority of hospital buildings would be in compliance with the Act's standards within 25 years. However, hospital buildings were not, and are not, being replaced at that anticipated rate. In fact, the great majority of the State's urgent-care facilities are now more than 40 years old" (California Seismic Safety Commission, 2001: 3).

5.2.3.3 The Regional Level: An Example of Infrastructure

Since the 1970s, ABAG has been working on various comprehensive regional plans, including those relating to earthquakes and other hazard mitigation plans. For decades, the organization has been working closely with other regional agencies, and in close partnership with city and county governments. As it pertains to earthquakes, AGAG's goal is "to maintain and enhance a disaster-resistant region by reducing the potential loss of life, property damage, and environmental degradation from natural disasters, while accelerating economic recovery from those disasters" (Perkins & Hutchings, 2010: 2). As one longtime specialist of Bay Area disaster preparedness explained, the organization of the committee was specifically created to allow for an overlap between actors, in order to have information circulate more fluidly between all those involved in disaster mitigation:

> The two ABAG committees are meeting jointly [...]. One [is] called Lifeline Infrastructure and Hazards Committee, which consists of representatives from transportation groups, water people, people who work for water districts, [and] there's a couple emergency managers that also sit on the committee. The other one is called Earthquake and Hazards Outreach Committee, and [it] deals with housing public business outreach and so on. We have residential retrofit contractors, structural engineers who deal with building, [and] also have a couple emergency managers [...]. Not that much overlap, those are my two specialty committees. There's the Regional Airport Planning Committee which is the joint committee of ABAG and MTC [or Metropolitan Transportation Commission] and BCDC [the Bay Area Conservation Development Commission]. They deal with the airport issue. If we get this Caltrans [California Department of Transportation] and airport, it will go to the Lifeline Committee. If we get money from PG & E to look at the housing soft-story[43] issue, that will be dealt with by [the] Earthquake and Hazards Outreach Committee. If we get money to do disaster recovery, that's more RPC, the Regional Planning Committee. RPC is 50 percent elected officials, high-level people in local government. They're different people, but there's a little overlap,

[43] "A soft story building is a multi-story building in which one or more floors have windows, wide doors, large unobstructed commercial spaces, or other openings in places where a shear wall would normally be required for stability as a matter of earthquake engineering design. A typical soft story building is an apartment building of three or more stories located over a ground level with large openings, such as a parking garage or series of retail businesses with large windows" (Wikipedia Contributors, 2019).

> you need some overlap so you have people who say," Oh, I was there, this is what happened." It's not just the staff person saying," Oh, that committee said this"; they can say, "Oh, I chaired that committee." That's why we have elected officials chairing committees. When I go down and have these meetings because we're writing these recovery plans for SF and Oakland, those are completely different people with a few exceptions. The person that's reviewing the housing section in San Jose for the San Jose Committee [...] is the same person that sits on Earthquake Hazard. [J.8]

This circulation of information is important. ABAG and the various local governments of the Bay Area follow the recommendation of the Stafford Act,[44] which "defines mitigation as 'any sustained action taken to reduce or eliminate the long-term risk to human life and property from hazards.'[45] As mitigation activities are undertaken, the risks associated with disasters decrease" (Perkins & Hutchings, 2010). Consequently, the ABAG framework for research and its recommendations of policies has followed two different, but complementary, perspectives on how to deal with disasters[46]:

[44] "The Stafford Act is a 1988 amended version of the Disaster Relief Act of 1974. It created the system in place today by which a presidential disaster declaration of an emergency triggers financial and physical assistance through the Federal Emergency Management Agency (FEMA). The Act gives FEMA the responsibility for coordinating government-wide relief efforts. The Federal Response Plan implements include the contributions of 28 federal agencies and non-governmental organizations, such as the American Red Cross. It is named for Sen. Robert Stafford (years in Senate 1971–1989), who helped pass the law. Congress amended it by passing the Disaster Mitigation Act of 2000, and again in 2006 with the Pets Evacuation and Transportation Standards Act." (Wikipedia Contributors, 2018a).

[45] Source: 44 CFR, Section 201.2, pertaining to Section 322 of the Stafford Act, 42 U.S.C. 5165.

[46] As discussed in the Introduction, I use the ABAG definition for a "disaster" as the following: "A disaster is a natural or man-made emergency whose response needs exceed available resources. When local government resources are exceeded, the California Governor's Office of Emergency Services (State OES) is contacted and the Governor is requested to declare a State Disaster. When State resources are exceeded, State OES contacts the U.S. Department of Homeland Security's FEMA and the President is requested to declare a National Disaster. This Presidential Declaration triggers funding resources for the public, the state, and local governments to use for clean-up, repair, recovery, and mitigation" (Perkins, 2005).

1. Increase emergency response capability.
2. Prevent or lessen the impacts of future incidents, reducing the need for larger and larger response capability. (Perkins & Hutchings, 2010: 1).

Dealing with disasters needs to take into consideration not only the hazards in relation to just one element, but also the strong interdependencies of all actants.

Lifelines and infrastructure tend to be situated in the same location. Water, sewers, gas pipelines, communications and electrical cable are buried under local roads. A rupture in a high-density, extremely interconnected system can rapidly lead a cascade effect. Lifeline systems are also connected through a system of dependencies: electricity is heavily used in the transportation of water, the capacity to deliver fuel can impact back-up generators for critical operations facilities, and so on. Acknowledging this interconnection is acknowledging the multiple actors that will come into play: local governments, private companies, local authorities.

Cities, counties, and infrastructure providers should be working together but fortunately, many of the recommendation strategies made by ABAG remain underfunded (Perkins & Hutching, 2010).

5.2.3.4 The Local Level: The Example of Soft-Story Buildings

In April 2013, on the highly symbolic anniversary of the 1906 earthquake, the San Francisco mayor finally signed the long-awaited, mandatory, soft-story retrofit program, a law that forces homeowners of the 2800 buildings considered structurally unsafe in the event of an earthquake to reassess their properties. This measure was designed to have a direct impact on the resilience capacities of the 58,000 people and 2000 businesses currently using these buildings.

> Soft-stories buildings are the most common structures in San Francisco, where the apartments buildings have large, open spaces reserved for storefronts or garages on the ground floor. Consistently, these buildings have been pointed out as responsible for major damage and casualties during previous earthquakes. As one engineer stated: [the] typical San Francisco building would have parking on the ground level, [...] so mostly what you saw [with previous earthquakes] was [that] the ground floor collapsed over, and the three or four stories above pretty much come straight down. [J.12]

For a decade, together with unreinforced masonry buildings—brick buildings that have not been braced—soft-story buildings have been the major preoccupation of structural engineers and urban analysts. A researcher who worked on the building assessment for a regional planning agency noted:

> 16,000 housing units fell down in the Loma Prieta earthquake, they were older homes that were not bolted to their foundation or had crawlspaces where the outside wall was not properly braced. [J.8]

As the following expert recalled, organizations such as ABAG or SPUR work as facilitators, and help push for good practices on official committees:

> There is an organization called SPUR. They have a couple of meetings but they also bring people that have common interests together. So public work structure for small cities will work together and they will share information. Members of city councils will work together; I think that is a very important organization for diffusion policies from city to city. When a city like San Francisco is addressing the earthquake safety of small residential buildings, the members of the SF Board of supervisors will sit down with members of the city council of other cities [to discuss the issues involved ...]; that is how those ideas [are] diffused. [T.4]

Here, again, the objective is to build resilience through data collection. ABAG estimates that across the San Francisco Bay Area, 139,765 units would probably collapse in the event of a major earthquake, or, at the very least, would not meet the "shelter in place standard". Active mobilization through committees has been earthquake experts' answer to the apparent previous lack of interest in earthquake safety. Through information and advocacy, they have been able to "mobilize public opinion or the political process in pursuit of their goal" (Geschwind, 2001).

After the signature, in 2013, of the Soft-Story Ordinance Bond,[47] the San Francisco Mayor stated:

[47] Mayor Edwin M. Lee signed the Mandatory Soft Story Ordinance, San Francisco's recently approved Ordinance No. 66-13 which requires the retrofitting of all buildings with the following characteristics: Wood-frame structures; those containing five or more dwelling units; those having three or more stories; and those permitted to be constructed prior to January 1, 1978.

> In order to be a truly resilient City, we must protect our residents and make sure their homes are safe after a major seismic event. This mandatory seismic retrofit program will protect San Franciscans, protect our housing stock and ensure San Francisco can rapidly recover from the next earthquake. Today, we renew our commitment to making sure that disasters such as the 1906 Earthquake and Fire do not devastate our City again. (Mayor Lee & Supervisors Sign City's Mandatory Seismic Retrofit Program For Soft Story Buildings, 2013)

But with more than 55,000 residents living in these soft-story buildings, the road is still long before all San Franciscans can live in a safe house and, if woken up in the middle of the night by the next Big One, share James' enthusiasm: "By Jove, here's [this] old earthquake, after all! And a jolly good one it is, too!" (James, 1906).

5.3 A Transition Still to Be Built: Between Science, Expertise, and Public Mobilization

Even as it becomes more and more complex, seismology has remained a science of observation, and prevention has always been a central preoccupation, as the need for interdisciplinary work, translation, and connection was acknowledged early on. Indeed, questions of resilience and vulnerability. For instance, after the Great San Francisco Earthquake in 1906, Andrew Lawson had already, pointed out the vulnerability of the existing and future buildings in the Marina District, a San Francisco neighborhood built on soft sedimentary soils.

This specific area would eventually be badly damaged, eight decades later, during 1989 Loma Prieta earthquake. Despite this considerable step,this revolution, some would say, at the time of the Lawson Report, seismology was still a young science. It was not until the middle of the twentieth century—and after 50 years of controversy—that the development of tectonophysics finally allowed the confirmation of the theory of continental drift and the definitions of specific seismic zones. Today, the progress of seismology as a science have been considerable, but it is still difficult to hear the voices of the scientists who made that possible and who are advocating for more residential measures in the political arena.

In 1964, an M.9.2 earthquake hit the state of Alaska. It caused considerable damage and 139 casualties, and attracted the interest of several UC Berkeley faculty members from the department of civil engineering, planning, and architecture who started thinking about what damage an earthquake of this dimension could cause in the Bay Area. In 1968, a monograph, *Earthquake Hazard in the San Francisco Bay Area: A Continuing Problem in Public Policy* (Steinbrugge, 1968) looked at potential consequences of a similar earthquake occurring in the Bay Area. The book drew the attention of California senator Alfred Alquist. It is said that the Joint Committee on Seismic Safety for the State of California was born from a dinner at the Faculty Club on the Berkeley campus (Chakos, Schulz, & Tobin, 2002).

Three years later, in 1971, after a devastating earthquake hit the San Fernando Valley, Alquist introduced the Hospital Seismic Safety Act, which passed in 1972, and the Alquist-Priolo Earthquake Fault Zoning Act, which prohibits "the siting of most structures for human occupancy across traces of active faults that constitute a potential hazard to structures from surface faulting or fault creep" ("Alquist-Priolo Earthquake Fault Zoning (AP) Act," 2012). The year 1972 was also when California's Governor Ronald Reagan established the Governor's Earthquake Council, which was then followed, in 1974, by the Seismic Safety Act—the bill that created the Seismic Safety Commission.

Once these first connections between the sciences and policy-makers were established, questions were raised on two grounds:

- What are the earthquake predictions for southern and northern California?
- What should be done to avoid casualties and damage?

The earthquake problem was taken seriously in the following decades but, as no serious earthquake occurred, concern began to fade away. The 1970s were a very active period, as R.28 recalled:

> Little known and less remembered was the initiative of Jerry Brown, then in his first two terms as Governor in the late 1970s and early 1980s in the creation of the Governor's Earthquake Taskforce comprised of 400 "experts" on 40 committees that were tasked to assess the risk and response capacity of the state and local governments and to develop plans to fill the gaps. The Governor directed the State Geologist, James Davis, to create scenarios for the most likely earthquakes in southern and northern California that would enable the Task Force committees to focus on the impacts of the earthquakes. These (eventually 7+) scenarios were

> pioneering efforts to quantify the impacts of the ground motions on structures, lifelines, infrastructure, hospitals, schools etc. and painted a picture story of what an earthquake like 1857 or 1906 would be like in today's physical environment. The Task Force went away but the scenarios remained and became the basis of planning and advocacy for two decades, when they were superseded by more detailed scenarios produced by HAZUS. (I served to represent government emergency management on the committee). (Personal communication, 10.03.13)

However, things changed drastically in 1989, when the Loma Prieta earthquake hit the Bay Area and killed 63 people. Fear of natural disasters then spread even further two years later, when the Oakland Fire destroyed 3500 single-family homes and killed 25 people. Suddenly, "the possible" had taken concrete form; the true-life scenarios were real and close—too close. These two events generated a great number of stories, and high levels of sorrow, concern, and attention.

The Bay Area was in a state of shock and residents felt the full sense of the fragility of life. These events also brought to the front lines a new generation of concerned citizens—average lay people—who wanted to know if, and how, their communities and their city would be able to face future earthquakes or other major natural disasters.

The early 1990s in the Bay Area are what people in the field of disaster studies call "a window of opportunity." Community organizing was big after the Oakland Fire, and a great deal needed to be done at every level, from the individual up to the state. Eventually, the residents of this largely educated, liberal neighborhood did what it took to improve their resilience, and to avoid future, similar dramatic events. In the city of Berkeley itself, people gathered together and started to ask questions, make phone calls, and exhume reports and documents long forgotten by city officials. Through their efforts, they quickly made solid, long-term relationships with renowned experts in the field of seismic safety.

The mayor of Berkeley at the time had long been an advocate for seismic safety, and brought together many experts and laypeople who were interested in the topic. Integrating a new profile, the expert community pursued its advocacy of seismic safety. In just one decade, Berkeley residents voted in several measures that, taken together, allowed more than $350 million from local taxes to be used for municipal safety improvements, including seismic retrofitting of city buildings and various school safety programs (Chakos et al., 2002). Some have said it was one of the most important "self-inflicted" increases of local taxes in the history of the USA.

These actions were made possible by the social and political context in which they occurred. For several decades, Berkeley, and the Bay Area at large, had been known for its strong attachment to a lively democratic tradition: in this "Atmosphere in which leaders take responsibility for reducing risk" (Chakos et al., 2002), decision makers took the time to integrate recommendations from local communities leaders, social scientists, engineers and seismologists.

This integration of knowledge also lead to challenge the divide between experts and lay people. Interested in what happen during a disaster, researchers started collecting data to question the assumptions—frequent in some risk management circles—that residents were uneducated and irrational. After Loma Pieta earthquake UCLA's Institute for Social Sciences Research (Bourque & Russell, 1994) showed that: "there was little evidence that people panicked [during the earthquake]. Most moved to a protected location or stayed in place. People at school and work were more likely to take action to protect themselves."

In the same period, others researchers were debunking the myth of the "irrational" resident. The Pacific Earthquake Engineering Research Center explained that:

> Resistance to implementation of earthquake ordinances by individual homeowners may not be irrational, but merely due to a decision frame that is different from those of an economist or engineer. Understanding the decision frames of people who eventually have to pay the cost of the regulations, and providing appropriate incentives for implementation should therefore be an important part of both regulatory and economic analysis. (Von Winterfieldt et al., 2000)

An other interesting example of this change of focus revolve around the interpretation, and trust, of earthquake predictions. Back in the 1970s, social sciences researchers have tried to evaluate the public's response to earthquake predictions. Results of this research consistently showed that residents did not pay much attention to them and, of those that did, only a few took the step and get prepared. Researchers concluded that people did not fully understand earthquake risk: "public perception of seismic hazard rarely conform[s to] common sense," one study from 1982 stated (quoted by Coen, 2013: 273), questioning—in an interesting, rhetorical manner—the definition of "common sense".

In the Bay Area, the case for residents' lack of "common sense" was carried further by Aaron Wildavsky (1930–1993), Professor of Political Science at UC Berkeley. A specialist in public policy and budget theory, Wildavsky was also co-author, along with Mary Douglas, of *Risk and Culture* (Douglas & Wildavsky, 1983). During his career, Wildavsky had trained and mentored many researchers who later became prominent professionals working in the field of risk and disaster studies. During a discussion with one of his former students, she explained:

> Aaron Wildavsky was my mentor; he was a very respected and very kind person. However, like many of his students and colleagues from the first years of his career, I did not follow him when he took a very reactionary turn in the 1980s and the 1990s. [L.30]

Indeed, during the 1980s Wildavsky became a controversial figure when he became associated with the "Merchants of Doubt," a community of scientists recruited by the tobacco industry to muddy discussions regarding the potential carcinogenic risk of cigarettes[48] (Oreskes & Conway, 2011) Berkeley.[49] In the case of earthquake risk, while drawing conclusions about the public's heuristic biases and failures of understanding, these lines of research missed the opportunity to question the fact that most earthquake predictions were proven to be inaccurate (Mileti & DeRouen Darlington, 1995).

[48]Wildavsky was a supporter the Advancement of Sound Science Coalition, a "front" group established by the company Philip Morris, which was a consultancy company for the tobacco industry in 1979.

[49]"Prospect Theory is a behavioral economic theory that describes the way people choose between probabilistic alternatives that involve risk, where the probabilities of outcomes are known. The theory states that people make decisions based on the potential value of losses and gains rather than the final outcome, and that people evaluate these losses and gains using certain heuristics" (Wikipedia Contributors, 2019) It is interesting to note that Kahneman and Tversky's Prospect Theory has somehow escaped the control of its authors; a recent commentator noted, "it is important to emphasize that the goal of the heuristics-and-biases literature is emphatically not to show that people are fools, or that they are systematically irrational. On the contrary, Kahneman and Tversky emphasized that the relevant heuristics are efficient and generally work well." But in the laboratory, at least, people who use the heuristics sometimes blunder, and it is the blundering that has attracted the most academic attention (Sunstein, 2002).

Indeed, the development of an instruments-based earthquake science was considered too complex to make intelligible "many of the scientists and the engineers who began to worry about seismic hazards in the 1970s had little hope of explaining the risk to the public. Seismic Risk was defined as a cost benefit calculation made by seismologists and engineers," as Coen wrote. As a consequence, Coen recalled, the researchers recommended "vigorous public information and education campaigns [...] that eliminate the biases introduced when the public interprets information" (Coen, 2013: 273).

Thus, when the Parkfield Experiment began, the Director of the U.S. Geological Survey very publicly made the prediction of an earthquake on April 5, 1985. As one researcher recalled:

> The statement forecasted an earthquake of magnitude 5.5 to 6.0 in the next several years (1985–1993) with more than a 90% probability that it would happen, and the forecast also stated that there was potential for this quake to be magnitude 7. The release of this prediction was a major media event, and California's Office of Emergency Services eventually prepared and mailed a brochure describing the prediction and recommended action to the more than 122,000 central California households within the extended area at risk, assuming a magnitude 7. (Mileti & DeRouen Darlington, 1995)

Researchers found that the public's response to the prediction was mostly positive, and that "no negative societal impacts were observed" (Mileti & DeRouen Darlington, 1995). Later, the USGS, who had already spent nearly $19 million on this project, went so far as to issue a public warning that a powerful earthquake could hit on a specific day. On October 21, 1992, Parkfield's local newspaper, *The Post and Courier*, stated that, "prompted by a 4.7-magnitude quake Monday night, the State Office of Emergency Services warned residents in a 30-miles radius of a 1 in 3 chance of a large earthquake by Thursday night," (The Post and Courrier with A.P., 1992) and also noted that this was the only time the USGS "has issued a formal earthquake prediction specifying time and magnitude" (The Post and Courrier with A.P., 1992).

On the day of the expected earthquake, the reporter sent to cover the story could only state that, "the warning did not deter residents from their routines, despite a wave of reporters and televisions crews". The quotidian activities of this rural town remained the same: "one woman hosed down the yard outside her trailer. Another made soup for a family. A farmer drove his tractor along the road" (The Post and Courrier with A.P., 1992). As we now know, no earthquake occurred that day, or any other day, up until September 28, 2004.

The Parkfield Earthquake was a non-event, yet it still helps to create awareness. Despite the difficulties in instauring earthquake risk with the help of instruments, the National Earthquake Hazards Reduction Program launched a series of community programs to increase the public's awareness of possible risks; these turned out to be far more successful than had been expected. Starting with the Southern California Earthquake Preparedness Project, a small group of nine people came together to begin to think about how to translate the fragile body of scientific data into efficient public policies, emergency response systems, and community resilience.

Together, FEMA and the State of California founded a project that was built on the theoretical framework of human behavioral change. With this, a great deal of money was then invested to push for what was considered a necessary step to reduce city vulnerabilities in California. Three major groups of stakeholders were identified: local governments, businesses, and the general public. To reach out to these stakeholders, several products were developed, one of which was a book entitled *Marketing Earthquake Preparedness* (Brady & Eisner, 1986), that provided a useful script for local governments and municipalities to help get the word out.

One of the experts involved in preparing Parkfield residents for the possible earthquake was my interviewee R. An architect by training, R. had studied at Berkeley in the 1960s and, during his years there, he had been inspired by the work of Saul Alinsky, a Chicago native and community organizer, who had written the book *Rules for Radicals* (Alinsky, 1989). First active in the labor movement, Alinsky had worked in some of Chicago's poorest neighborhoods in the late 1930s and had rapidly become an inspiration for activists and politicians alike.[50] While

[50] Inspired, but also criticized, Alinsky's followers redefined the objectives and methodologies of community organizations and mobilization. In an article for *Illinois Issues* in 1988, Barack Obama wrote: "In theory, community organizing provides a way to merge various strategies for neighborhood empowerment. Organizing begins with the premise that (1) the problems facing inner-city communities do not result from a lack of effective solutions, but from a lack of power to implement these solutions; (2) that the only way for communities to build long-term power is by organizing people and money around a common vision; and (3) that a viable organization can only be achieved if a broadly based indigenous

preparing southern California for a potential earthquake, the advocacy and community organization background of this former radical student, now a professional, came into play. As R. explained, Alinsky's work made a difference:

> We wanted to empower people. We wanted them to feel they have the control of the situation. We would go from city to city and talk to the building department people, show them our scripts and propose to talk to their supervisor or to city council. We offered them a "planning partnership" which helped them identify the buildings, infrastructures and systems at risk and what could be done to be improve their organization and their capacity to face a disaster. [R.29]

Focusing on a very different approach than those favored at the time, this team brought together scientists and senior political staff to explain what an earthquake would look like for each institution that they worked with, including: "a Southern California County (San Bernardino County), a large city (Los Angeles), a small city (Westminster in Orange County), and a corporation (Security Pacific Bank)" (Geschwind, 2001: 217).

As R. recalled, these meetings included a great deal of coffee, and a great deal of listening, even more so than talking. Their motto was "do not focus on *you should*," but, rather, help people and institutions understand what could happen and how they could take steps to mitigate risks. Most importantly, they provided a safe environment to think about the "unthinkable"—the injuries, deaths, and financial devastation that a large catastrophe could provoke. This group's focus was to act as a translator for the government people and private institutions, "making sense of the language of engineers" (Chakos et al., 2002: 7), with a very strategic perspective:

> For each of these institutions, project staff members worked together with emergency preparedness personnel to develop detailed response plans and contingency plans for a large earthquake. Once these prototype plans had

leadership—and not one or two charismatic leaders—can knit together the diverse interests of their local institutions. This means bringing together churches, block clubs, parent groups and any other institutions in a given community to pay dues, hire organizers, conduct research, develop leadership, hold rallies and education campaigns, and begin drawing up plans on a whole range of issues—jobs, education, crime, etc." (Obama, 2008).

> been worked out, they could then be transferred to similar entities across the region. In this manner, earthquake planning increased significantly in Southern California, because individual preparedness offices now had detailed models of emergency plans [...] available to them. (Geschwind, 2001: 217)

Within just a few years, the team was involved in shaping earthquake preparedness plans for San Francisco, Berkeley, Oakland, Santa Cruz, Palo Alto, and Santa Barbara. In each of these cities, they focused on what they termed "sweet spots"—places and institutions that were supposed to be protected by law, and which would accommodate vulnerable populations including schools, hospitals, and government buildings. The project was so successful that it was next reproduced in other parts of California.

Thanks to the domino effect, as the work of this small team reached out to more communities, many cities started to take concrete, visible actions. In one article, three experts—two of whom were respondents for this research—recalled how proactive the city of Berkeley, and its residents, had been when approached regarding the question of seismic safety:

> In four local elections Berkeley voters have approved over $390 million in local taxes to fund mitigation projects. In addition, the City now rebates 1/3 of its real estate transfer tax, up to a maximum of $1500, for seismic retrofit—Goldfarb was the swing vote in agreeing in 1991 to increase the transfer tax only if the Council agreed to the rebate. As a result, over 39% of Berkeley's 22,000 single-family residences and over 30% of small multi-family buildings now have improved seismic resistance, at a cost to the City of over $10 million in foregone taxes. The Disaster Council—like the Seismic Safety Commission at the state level—has kept earthquake safety on the agenda of the City Council. (Chakos et al., 2002)

In addition, these campaigns, the implementations of prevention, and the educational programs of the late 1990s have also engaged public workers with concrete aspects of earthquake risk—the safety of their office buildings, their responsibility vis-à-vis the public, and so forth. In so doing, they have helped build awareness by making the materiality of earthquake risks visible. The following respondent, a librarian for the city of Berkeley, benefited from these programs; she recalled:

> What I saw in Berkeley was much more about prevention and how to respond to disasters that came out of that situation. Where I work, they started having regular sessions, not just drills, but they take everybody through the building, and show them where all the emergency exits are, what to do in case of an earthquake or a fire, what your responsibilities about getting people out of the building, how to deal with people in wheelchairs and that kind of thing. I don't remember having anything like that before the 1989 earthquake. [A.-M.1]

Finally, following the 1989 earthquake, large public and infrastructure works began, which made the potential risks even more visible for residents of the Bay Area. As one respondent commented:

> The concern that the Bay Area's residents had since the Loma Prieta earthquake within infrastructure is visible. They've seen the signs, gone through the disruption, getting the toll bridges and interchanges on highway bridges retrofitted. Their bills for water [were] increased to pay for upgrades to the water system. They've seen the fares and tolls and their property checks and bills change to pay for the retrofitting of the BART system. They've seen [that] this not only hit them personally, in terms of money, but they've also seen the construction happening. I think there is a larger group of people who understand they have to retrofit their homes than used to be the case. In terms of non-infrastructure, the things that have to be done by the individual to their own home, I don't think this is obvious. [J.8]

The considerable work of education and outreach risk relied on the idea that tailored recommendations gleaned from constant field work create better preparedness. Outreach and education have been progressively included infrastructure programms. The progressive instauration of a network of concerned experts has allowed for the development of risk awareness. Grounded in the experience of waiting for the next big earthquake, this community of Earthquake Junkies has succeeded in making disaster mitigation regulations a major political subject over the last few decades.

Does this mean that the Bay Area is now ready for the next Big One? The answer to this question is, of course, probably not. In the late '90s, the team of nine, then called the Bay Area Regional Earthquake Preparedness Project (BAREPP), was integrated into the Governor's Office of Emergency Services, and, according to R., who was then managing this entity, it slowly but surely lost a certain degree of its freedom because of the political directions given by then-Governor Gavin. Not being visible in the media was detrimental for this group which, for decades, had been a very vocal advocate of the disaster preparedness community. Soon enough, one of the ricochet effects of the events of 9/11 was a dismantling of previous disaster preparedness, as well as the responsibilities of responders for certain organizations nationwide. This included the field of earthquake preparedness: "It was an overnight change," recalled R. Within months, BAREPP's library in Oakland was shut down, and twenty years of documentation and archives were thrown away. In 2003, California Governor Arnold Schwarzenegger took office: "he said to the people of California 'If you need something call me!' which doubled the amount of work" [R.29]. As R. recalled, this call largely disturbed the organization of the group, forcing them to answer any kind of call, not all related to earthquake preparedness. The expertise and dedication of the BAREPP was no longer considered capital and was diluted to fit the personal political agenda of the California governors.

As the earthquake became a very low priority in the political arena, for a decade or more the message from local governments was, essentially, "You're on your own." Experts worried about a total disengagement of California and federal government support. During this time, they were hoping that the actions taken during the '90s would be enough to limit the effects of major catastrophes, including any large-scale earthquakes (Chakos, 2006).

Today, after 30 years of working in the field of disaster preparedness in the Bay Area, R. has retired, and the work of information and advocacy is pursued by the new generation of Earthquake Junkies, working with dedication to maintain the level of awareness necessary to good earthquake preparedness. Large, highly technical projects, such as the early -warning earthquake system conceived by UCB Seismology Lab, are still pursuing to the effort. But as R., the scientist leading the project, stated, "In fact our project has very little, very little of the social part of the question." [R.9]. Another scientist added that outreach and public prevention remains poor relative to action, and thus, for many different reasons:

> All these things, it is all political. (A) We were never trained to do this, so we're not that good at it. (B) Frankly we don't have that much time to

> do it, we're still doing research, going and meeting with all these groups and educating people takes a lot of time. Having people who could go and work with these different groups and interact with different groups, that would be great. But again, that comes back to money and finding a way to do that. That would be the primary way that I think this could be improved, by having persons who could engage all of the groups and their job would simply be to work, [to] build the collaboration. [J.15]

To answer those critics, some of the Earthquake Junkies want to focus on what *has* been accomplished, and what a new generation of Earthquake Junkies is still out to accomplish. In 2011, in an article called "Getting a Jump on the Next Big Disaster," Chakos wrote, "The best disaster preparedness in the world is happening in San Francisco, which has quietly developed public-private partnerships that can swing into action and save lives when and if the predicted 'Big One' hits" (Chakos, 2010).

In this piece, Chakos argues that the Earthquake Junkies are still in and have learned their "ABK," (All but Katrina) an engage in multifaceted collaborations.

> Elected officials debate mandatory safety improvements for apartment buildings that could collapse in an earthquake. San Francisco risk managers and their insurers map disaster-related financial recovery plans. Earthquake engineers strategize with regional utility companies about how to get the lights back on, the water flowing, and transportation moving after the next disaster. Youth leaders connect area residents to their web 2.0 neighborhoods. S.F. State University student organizers survey the Excelsior and Castro neighborhoods on disaster-readiness, part of the Neighborhood Empowerment Network program". (Chakos, 2010)

Taking the opposite tack of what Mike Davis recommended in his famous article "The case for letting Malibu Burn" (Davis, 1998), San Francisco Bay Area Earthquake Junkies have decided to work forcefully against letting San Francisco and the Bay Area be wiped off the map during the next earthquake. Working in committees, finding consensus, these experts provide and promote the best earthquake science and preparation possible. In so doing, they have worked "with" the earthquake, understood a complex phenomenon. More than a century after William James account the multiple dimensions of the "perception of the [earthquake] as a living agent" and its slow instauration among the San

Francisco Bay Area occupants is what Earthquake Junkies continue to explore as they await the next Big One.

Waiting for the Big One is a collective endeavor that, San Francisco has mobilized concerned scientists, experts, and citizen. As I was approaching the end of this research, the discussions with R.28 had, in a way that was maybe even more strongly emphasized, demonstrated the strength of this attachment to the idea of preparedness that is carried and transported through time and space. R.28 was introduced to me through a common connection who mentioned that he was very involved in the Disaster Risk reduction of the Bay Area, that he had worked for several non-profit and state agencies, and that he had achieved important connections with Japanese experts over the years. He also mentioned that R.28 was "very much a Bay Area resident, and advocate" (Personal communication, 09.4.13).

When I met R.28, he had decided to "really" retire, which for him means to try to let go, as much as possible what had been a life of work and dedication, often interrupted with a considerable amount of frustration and deception, coming both from the scientific and political world. R.28 trained as an architect during the Free Speech Movement had been influenced by the work of Alinsky. As he recalled, his stepping into the field of earthquake preparedness had a great deal to do with the impact that the previous generation of experts had had on him:

> California was "blessed" with engineers and geologists, not just in the Bay Area, and created the modern science of seismology and structural engineering at Caltech, Berkeley and Stanford. These were bigger than life people that included John Blume at Stanford, Henry Degenkolb, a Structural Engineer in SF, Karl Steinbrugge a Structural Engineer who worked for the Insurance Services Organization and taught at the architectural school at Berkeley, Bob Olson a Political Scientist who headed the CA Seismic Safety Commission for its first 20 years, Henry Lagorio, as Architect on the Faculty of Berkeley who championed designing (architectural) for seismic forces, Bruce Bolt at the Berkeley Seismological Laboratory, Stanley Scott at Berkeley's Institute for Governmental Studies who pioneered seismic public policy initiatives, and the members of EERI in northern and southern California who created the seismic code provisions that eventually found their way into the Uniform Building Code. Steinbrugge, Lagorio and Scott were my mentors and got me interested studying earthquakes. It all happened here because there were mentors and leaders. (Personal communication, R. 10.03.13)

For more than three decades, he became one of the Earthquake Junkies, participating in multiple programs and initiatives from the Parkfield experiment to the development and management of the Bay Area Regional Earthquake Preparedness Project, working as Regional Administrator for the California Governor's Office of Emergency Services, and being appointed Visiting Scholar and Professor at the Center for Disaster Reduction Systems, Disaster Prevention Research Institute, in a Japanese University. During his active years, he also managed the California Integrated Seismic Network, a state-funded dense urban seismic network in northern and southern California, through funding to the California Institute of Technology, California Geological Survey, and the University of California Berkeley Seismological Laboratory, used for research and predictions, and also worked in several non-governmental organizations and FEMA. In addition, he co-wrote books and articles that mainly focus on the best way to engage with a larger public and the development of material that can be helpful in developing better preparedness.

At the time of our discussion, he was working hard to identify the next generation of Earthquake Junkie researchers and experts with whom he could share his knowledge and to whom he could pass on the challenge that had been his for decades. As we were reflecting on his trajectory, he wrote: "My career in community work, teaching, research and trying to fit into the state bureaucracy seemed today like 35 years of chaos. The logic only appears in hindsight." (Personal communication, 09.25.13). I hope through this work to have contributed to giving visibility to this collective experience and inscribed in a larger effort to make San Francisco a place "of action, where all are concerned together," as William James wrote in 1906. He continued: "at San Francisco the need will continue to be awful, and there will doubtless be a crop of nervous wrecks before the weeks and months are over, but meanwhile the commonest men, simply because they are men, will go on, singly and collectively, showing this admirable fortitude of temper" (James, 1906).

References

ABAG. (2013a). *Resilience initiative—Building a disaster resilient Bay Area*. Retrieved from http://quake.abag.ca.gov/projects/resilience_initiative/.

ABAG. (2013b). *Soft story buildings*. Retrieved from http://quake.abag.ca.gov/housing/softstory/.

Alinsky, S. (1989). *Rules for radicals*. New York: Vintage; a edition.
Alquist-Priolo Earthquake Fault Zoning (AP) Act. (2012). Retrieved from http://www.conservation.ca.gov/cgs/rghm/ap/Pages/index.aspx.
Bonowitz, D. (2009). *The dilemma of existing buildings: Private property, public risk*. Retrieved from http://www.spur.org/publications/library/report/dilemma-existing-buildings.
Bourque, L., & Russell, L. (1994). *Experiences during and responses to the Loma Prieta earthquake*. Sacramento: California Seismic Safety Commission.
Bradford, D. (2006). *Earthquake: The other insurance crisis*.
Brady, J., & Eisner, R. (1986). *Marketing earthquake preparedness: Community campaigns that get results*. Diane Publishing.
Burby, R., & May, P. (1998). *Making building codes an effective tool for earthquake hazard mitigation* (College of Urban and Public Affairs (CUPA)—Working Paper, 1991–2000, Paper 3). Retrieved from http://scholarworks.uno.edu/cgi/viewcontent.cgi?article=1002&context=cupa_wp.
California Seismic Safety Commission. (2001). *Findings and recommendations on hospital seismic safety*. Sacramento: State of California Seismic Safety Commission.
Callon, M. (1986). Some elements of a sociology of translation: Domestication of the scallops and the fishermen of Saint-Brieuc Bay. In J. Law (Ed.), *Power, action and belief: A new sociology of knowledge?* (pp. 196–223). London: Routledge.
Chakos, A. (2006). *Message to locals in California disasters: YOYO-You're on your own*.
Chakos, A. (2010). Getting a jump on the next big disaster. *Huffington Post*. Retrieved from http://www.huffingtonpost.com/arrietta-chakos/getting-a-jump-on-the-nex_b_588990.html?.
Chakos, A. (2011). *Toward a resilient future: A review of Palo Alto's emergency readiness*.
Chakos, A., Schulz, P., & Tobin, T. (2002). Making it work in Berkeley: Investing in community sustainability. *Natural Hazard Review, 3*, 55–67.
Coen, D. (2013). *The earthquake observers: Disaster science from Lisbon to Richter*. Chicago: The University of Chicago Press.
Davis, M. (1998). *Ecology of fear: Los Angeles and the imagination of disasters*. New York: Vintage Books.
Douglas, M., & Wildavsky, A. (1983). *Risk and culture: An essay and the selection of technological environmental dangers*. Berkeley: University of California Press.
Fante, J. (1939). *Ask the dust* (Modern Cla). New York: HarperPerenial.
Fehr, I. (2006). *The 1906 San Francisco earthquake and fire: Perspectives on a modern super cat*.

Filson, J., McCarthy, J., Ellsworth, W., & Zorback, M.-L. (2003). *The USGS earthquake hazard program in NEHRP—Investing in a safer future.* Retrieved from http://pubs.usgs.gov/fs/2003/fs017-03/.

Frankel, H. (2012). *The continental drift controversy: Evolution into plate tectonics* (Vol. 4). Cambridge: Cambridge University Press.

Geschwind, C.-H. (2001). *California earthquake, science, risk and the politic of hazard mitigation.* Baltimore and London: The Johns Hopkins University Press.

Grossi, P., & Zoback, M. L. (2009). *Catastrophe modeling and California earthquake risk: A 20-year perspective.*

Grossi, P., & Zoback, M. L. (2010). *1868 Hayward earthquake: 140 year retrospective.* Menlo Park: Risk Management Solutions.

Hansen, G. (1989). *Denial of disaster: The untold story and photographs of the San Francisco earthquake of 1906.* San Francisco: Cameron.

Heller, A. (2006). *Re-creating the 1906 San Fransisco earthquake.* Retrieved from https://www.llnl.gov/str/Sep06/Rodgers.html.

Hinman, E., & Hutchinson, D. A. (2005). *San Francisco's earthquake risk.*

Hutchings Mieler, D., & Brechwald, D. (2013). *Regional resilience initiative.*

James, W. (1906). *On some mental effects of the earthquake: The Youth's Companion.* Reprinted in James, H. Jr., 1911, *Memories and studies* (H. James, Jr.). Longmans, Green, & Co. (Vol. June). New York.

Kircher, C. A., Seligson, H. A., Bouabid, J., & Morrow, G. C. (2006). When the big one strikes again—Estinated losses due to a repeat of the 1906 San Francisco earthquake. *Earthquake Spectra, 22*(2), 297–339.

Lewis, J. (2008, October 21). *The coming quake: Is Los Angeles ready for the Big One?* High Country News. Retrieved from https://www.hcn.org/issues/40.19/the-coming-quake/?b_start:int=1.

Lynch, M. (2012). Self-exemplifying revolutions? Notes on Kuhn and Latour. *Social Studies of Science, 42*(3), 449–455. https://doi.org/10.1177/0306312712439120.

Maffei, J. (2010). *The coming Bay Area earthquake, 2010 update of scenario for a magnitude 7.0 on the Hayward fault.*

Main, I., Bell, A., Meredith, P., Geiger, S., & Touati, S. (2012). The dilatancy-diffusion hypothesis and earthquake predictability. *Geological Society of London, 367,* 215–230.

Mayor Lee & Supervisors Sign City's Mandatory Seismic Retrofit Program for Soft Story Buildings. (2013). San Francisco. Retrieved from Mayor Lee & Supervisors Sign City's Mandatory Seismic Retrofit Program for Soft Story Buildings.

Mileti, D. S., & DeRouen Darlington, J. (1995). Societal response to revised earthquake probabilities in the San Francisco Bay Area. *International Journal of Mass Emergencies Disasters, 13*(2), 119–145.

Min, L., & Perkins, J. B. (2008). *Long-Term Disaster Recovery Planning by Local Governments in the San Francisco Bay Area.*

November, V., Penelas, M., & Viot, P. (2009). When flood risk transforms a territory: The Lully effect. *Geography, 94,* 189–197.

Obama, B. (2008). *Why organize?* Retrieved from http://illinoisissues.uis.edu/archives/2008/09/whyorg.html.

Oreskes, N., & Conway, E. (2011). *Merchants of doubt: How a handful of scientists obscured the truth on issues from tobacco smoke to global warming* (Reprint edition). New York: Bloomsbury Press.

Paxton, J. (2004, August). Earthquakes: San Francisco at risk. *Spur NewsLetter.*

Perkins, J. B. (2005). *Taming natural disasters, multi-jurisdictional local government hazard mitigation plan for the San Francisco Bay Area.*

Perkins, J. B., Chakos, A., Olson, R. A., Tobin, L. T., & Turner, F. (2006). A retrospective on the 1906 earthquake's impact on Bay Area and California public policy. *Earthquake Spectra, 22*(S2), S237. https://doi.org/10.1193/1.2181527.

Perkins, J., & Hutchings, D. (2010). *Taming natural disasters: Multi-jurisdictional local hazard mitigation plan for the San Francisco Bay Area: 2010 update of 2005 plan.* Oakland.

Petersen, M. D., Frankel, A. D., Harmsen, S. C., Mueller, C. S., Haller, K. M., Wheeler, R. L., ... Luco, N. (2011). *Seismic-hazard maps for the conterminous United States, 2008.*

Poland, C. (2009). *The resilient city: Defining what San Francisco needs from its seismic mitigation policies.*

Reisner, M. (2003). *A dangerous place.* New York: Pantheon Books.

Society, S. H. (2006). Palo Alto, Stanford, and the 1906 earthquake. *Sandstone & Tile, 30*(1), 40.

Solnit, R. (2009). *A paradise built in hell: The extraordinary communities that arise in disaster.* London: Viking.

Souriau, E. (2009). *Les différents modes d'existence.* Paris: Presses Universitaires de France.

SPUR. (2013). *On solid ground.* San Francisco. Retrieved from http://www.spur.org/sites/default/files/publications_pdfs/SPUR_On_Solid_Ground.pdf.

Stallings, R. A. (1995). *Promoting risk: Constructing the earthquake threat (social problems and social issues).* New York: Aldine de Gruyter.

Steinbrugge, K. (1968). *Earthquake hazard in the San Francisco Bay Area: A Continuing problem in public policy.* Berkeley: Institute of Governmental Studies, University of California.

Sunstein, C. (2002). *Hazardous heuristic.* The Law School of the University of Chicago, Public Law. Retrieved from http://www.law.uchicago.edu/files/files/33.crs_.hazardous.pdf.

Thatcher, W., Ward, P., Wald, D., Hendley, J. I., & Stauffer, P. (2001). *When will the next great quake strike Northern California?*

The Post and Courrier with A.P. (1992). Town gets quake warning. *The Post and Courier*. Retrieved from http://news.google.com/newspapers?nid=2482&dat=19921021&id=wlZSAAAAIBAJ&sjid=1DYNAAAAIBAJ&pg=6194,1792815.

Tierney, K. J. (2000). *Public support and priorities for seismic rehabilitation in the East Bay region of Northern California* (No. #296).

Tobriner, S. (2006). *Bracing for disaster: Earthquake-resistant architecture and engineering in San Francisco, 1838–1933*. Berkeley: The Bancroft Library, University of California Berkeley.

Turner, F. (2004). Seventy years of the riley act and its effect on California's building stock. In *13th World Conference on Earthquake Engineering*. Vancouver.

U.S. Geological Survey. (2016). *When could the next large earthquake occur along the San Andreas fault?*

Von Winterfieldt, D., Roselund, N., & Kisuse, A. (2000). *Framing earthquake retrofitting decisions: The case of hillside home in Los Angeles*. Pacific Earthquake Engineering Research Center. Retrieved from http://peer.berkeley.edu/publications/peer_reports/reports_2000/0003.pdf.

Wikipedia Contributors. (2018a). Stafford Disaster Relief and Emergency Assistance Act. In *Wikipedia, The Free Encyclopedia*. Wikipedia, The Free Encyclopedia. Retrieved from https://en.wikipedia.org/w/index.php?title=Stafford_Disaster_Relief_and_Emergency_Assistance_Act&oldid=827132167.

Wikipedia Contributors. (2018b). Unreinforced masonry building. In *Wikipedia, The Free Encyclopedia*. Retrieved from http://en.wikipedia.org/wiki/Unreinforced_masonry_building

Wikipedia Contributors. (2019). Soft story building. Retrieved May 15, 2019, from https://en.wikipedia.org/w/index.php?title=Soft_story_building&oldid=885792427.

Winchester, S. (2011, March 13). The scariest earthquake is yet to come. *Newsweek Magazine*. Retrieved from http://www.thedailybeast.com/newsweek/2011/03/13/the-scariest-earthquake-is-yet-to-come.html.

Working Group on California Earthquake Probabilities. (1999). *Are Earthquakes really on the increase?*

Working Group on California Earthquake Probabilities. (2003). *Earthquake probabilities in the San Francisco Bay region: 2002–2031. Open-file report 03-214*. U.S. Geological Survey. Retrieved from https://pubs.usgs.gov/of/2003/of03-214/.

CHAPTER 6

Resilience as a Socio Technical Process

The process by which seismic provision, guidelines, regulations, and building codes have been defined and continue constantly to be revisited, is a good illustration of the progressive materialization of the network of awareness into pragmatic measures. It shows the long-term emergence of resilience in large urban areas.

Earthquakes in the twentieth century have prompted the introduction of regulatory frameworks. The Long Beach earthquake can be considered a starting point of a *modus operandi* that defined the collaboration between organizations dedicated to earthquake preparedness: joint committees working collaboratively and moving progressively from the local level to the state level, then up to the federal government. The M.6.4 Long Beach earthquake of March 10, 1933 had provoked important damages in South California's schools. Newspapers relayed the outrage of parents and Harry Wood, then director of the Seismological Laboratory at the California Institute for Technology, declared:

> The practical lesson of the recent temblor, as of all others, [...] is built well and choose or prepare strong foundations. Design for strength and [construct] conscientiously, using good materials. Avoid what experience has shown to be faulty. (Wood quoted in Geschwind, 2001: 107)

C. Mazel-Cabasse, *Waiting for the Big One*,
https://doi.org/10.1007/978-3-030-15289-5_6

In the following years, discussions were held which ultimately led to both the Field Act[1] and the Riley Act,[2] two laws that focused on the safety of public buildings and attempted to avoid proliferation of unreinforced masonry buildings.[3] Addressing what had become a common concern in California, the first seismic code, the Building Code of California, was published in 1939[4] after more than 10 years of work by small committees.[5] The expanding need for building regulations in the Bay Area in the 1940s[6] resulted from a real estate boom throughout the state.

[1]Assemblyman C. Don Field, a Republican, prepared the Field Act. For Turner, "The Field Act transferred the regulation of public school design and construction from local governments to the State's Division of Architecture. Its enforcement on over 70,000 school construction projects since then has generated significant improvements in the practice of earthquake engineering" (Turner, 2004: 4).

[2]Approximately 10 to 15% of California's buildings were constructed before 1933, at a time when few cities had building codes. The Riley Act "requires local governments to have building departments that issue permits for new construction and alterations to existing structures and conduct inspections. Permit fees paid by building owners generally fund the work of local building departments. The Act set minimum seismic safety requirements that have since been incorporated into all building codes." Western States Seismic Policy Council (http://www.wsspc.org/policy/California.shtml).

[3]"An unreinforced masonry building (or UMB, URM building) is a type of building where load-bearing walls, non-load-bearing walls or other structures, such as chimneys, are made of brick, cinderblock, tiles, adobe or other masonry material that is not braced by reinforcing beams. The term is used in earthquake engineering as a classification of certain structures for earthquake safety purposes and is subject to minor variation from place to place. URM structures are vulnerable to collapse in an earthquake. One problem is that most mortar used to hold bricks together is not strong enough. Additionally, masonry elements may "peel" from the building and fall onto occupants or passersby outside" (Wikipedia Contributors, 2018).

[4]A first edition of the Uniform Building Code was published in 1927. The eastern part of the United States has had building codes in place since 1914.

[5]It is interesting to note that committees remain the principal form of organization for structural engineers and the seismic community in general. Most of the reports produced in recent years have been done via committee consensus.

[6]The Vesano Building Code, named after Harry Vesano, former director of Public Works in San Francisco, was published in 1948.

6.1 Building for Resilient Infrastructures

Resilience is a slippery concept. Within a few years, it has overcome other concepts describing exposure to risk as well as individual, group, and infrastructure capacity to recover (Adger, 2000; Folke, 2006; Timmerman, 1981). Using the concept of resilience in lieu of other existing concepts has generated some epistemic tensions, as the polysemy of the term presents both an opportunity and a heuristic challenge (Mazel-Cabasse, 2017; Reghezza-Zitt, Rufat, & Djament-Tran, 2012). As a discursive and sometimes performative concept, resilience has allowed many important discussions on hermeneutics from both disaster and risk studies, but the multiple epistemological origins of the term opened the possibility of contradictory and sometimes incompatible interpretations.[7] Being or not being resilient encompasses many different realities, and to ordain a form of obligation to resilience might carry the possibility of stigmatization and value judgments (Bonanno, 2004; Jasanoff, 2014; Reghezza-Zitt & Rufat, 2015; Weichselgartner & Kelman, 2015). Therefore, resilience cannot be defined in advance, nor can it be requested. It can only be described after the facts, resisting the temptation of naturalization that reduces it to a simple function that would leave aside the dynamics and tensions constitutive of the social world (Comfort, Boin, & Demchak, Chris, 2010).[8]

Building resilient infrastructure has been the mission of Bay Area earthquake experts, scientists and concerned residents. Rob, an architect by training, was introduced to me through a common connection who referred to him as an "elder" because of his "old professional age" as well as his feeling of being "very much a Bay Area resident and

[7] "Resilience has been used by the physical sciences to describe the capacity of an object to bend under pressure and go back to its original form. It has been largely utilized by ecologists to define the capacity of a system to return to a state of equilibrium after a destructive event. These interpretations might prove useful for disaster management, emphasizing the positive side of vulnerability and providing a desirable future to lean toward" (Guarnieri, 2017).

[8] "Some cities do better in the face of disaster than others. It is tempting to describe apparent success in terms of resilience and apparent failure in terms of a shopping list of explanatory variables. Resilience then becomes the synonym for survival and the prescribed antidote for administrative shortcomings. This is too simple. ... Far from a fix-it-and-forget-it approach, resilience is the outcome of a long-term process; enduring resilience is a balancing act between risk and resources, between vulnerabilities and escalating or unmanageable risk" (Comfort et al. 2010: 272–273).

advocate" (personal communication, September 4, 2013) in the earthquake risk community. During more than three decades, Rob had been deeply involved in earthquake risk management, participating in multiple programs and initiatives, juggling positions at the local and regional levels, working with state agencies and non-governmental organizations alike, and teaching at international institutions. In his own account of this trajectory, he embraced a narrative that pays tribute to the importance of transgenerational transmission of knowledge and mitigation efforts. During our communication, he recalled his mentors:

> California was blessed with engineers and geologists, not just in the Bay Area, and created the modern science of seismology and structural engineering at Caltech, Berkeley, and Stanford. These were bigger-than-life people that included John Blume[9] at Stanford; Henry Degenkolb,[10] a structural engineer in San Francisco; Karl Steinbrugge,[11] a structural engineer who worked for the Insurance Services Organization and taught at the architectural school at Berkeley; Bob Olson,[12] a political scientist who headed the California Seismic Safety Commission for its first 20 years; Henry Lagorio,[13] an architect on the Berkeley faculty who championed

[9] John A. Blume (1909–2002) was considered by many of his peers as the father of earthquake engineering.

[10] Henry J. Degenkolb (1913–1989) served on the President's Task Force on Earthquake Hazard Reduction and received the American Society of Civil Engineers Ernest E. Howard Award.

[11] Karl V. Steinbrugge (1919–2001) was a professor of structural design at the University of California, Berkeley. One of his students recalled he had been very influential in defining earthquake risk: "For many of us in the design and construction fields, Karl Steinbrugge was the first person to explain the impact that earthquakes would have on our professional lives. He was quite passionate about the subject in clearly affirming that earthquakes needed to be better understood. In order to make improvements in engineering, planning, public policy, and emergency response, you had to learn from the past. As I think about the issues that we face today, I realize that what Karl was so passionate about has stayed with me long after the classes were over."

[12] Robert A. Olson, president of Robert Olson Associates, Inc., a consulting firm focusing on emergency management, hazard mitigation, community recovery planning, research, and knowledge transfer.

[13] Henry J. Lagorio (1923–2013), a professor of architecture and former dean of the College of Environmental Design, was known for "chasing earthquakes," his study of architectural design for earthquakes, and his loss-estimation methods.

> designing (architectural) for seismic forces; Bruce Bolt[14] at the Berkeley Seismological Laboratory; Stanley Scott[15] at Berkeley's Institute for Governmental Studies, who pioneered seismic public policy initiatives; and the members of EERI in northern and southern California who created the seismic code provisions that eventually found their way into the Uniform Building Code. Steinbrugge, Lagorio, and Scott were my mentors and got me interested in studying earthquakes. It all happened here because there were mentors and leaders. (Rob, personal communication, October 3, 2013)

This group of very interdisciplinary scholars and engineers had a set of common interests, but it is the M.9.2 earthquake that hit the state of Alaska in 1964, causing considerable physical damage and 139 casualties, that got them together. The event attracted the interest of Rob's mentors at the UC Berkeley. Soon, they started thinking about the kinds of damage an earthquake of this dimension could cause in the Bay Area; a monograph, *Earthquake Hazard in the San Francisco Bay Area: A Continuing Problem in Public Policy*, was published by Steinbrugge in 1968 (Steinbrugge, 1968). The book drew the attention of California senator Alfred Alquist. The Joint Committee on Seismic Safety for the State of California, "charged with investigating existing earthquake preparedness programs and with the development of a statewide seismic safety plan" ("Inventory of the Joint Committee on Seismic Safety Records," n.d.) and chaired by Alfred Alquist, was reportedly conceived during a dinner at Berkeley's Faculty Club the following year (Chakos, Schulz, & Tobin, 2002). Once these first connections between the scientists and policy-makers became established, it became important to know what were the chance of earthquake for southern and northern California and what should be done to avoid casualties and damage. The earthquake problem was taken seriously in the following decades and for earthquake experts and scientists, the 1970s were a very active period, as Rob recalled:

> Little known and less remembered was the initiative of Jerry Brown, then in his first two terms as governor in the late 1970s and early 1980s, in

[14] Bruce Bolt (1930–2005) was a professor of earth and planetary sciences at the University of California, Berkeley and a pioneer of engineering seismology.

[15] Stanley Scott was John A. Blume's student and later became a seismic-safety policy analyst with the Institute of Governmental Studies at the University of California, Berkeley.

> the creation of the Governor's Earthquake Taskforce—comprised of 400 "experts" on 40 committees that were tasked to assess the risk and response capacity of the state and local governments and to develop plans to fill the gaps. The governor directed the state geologist, James Davis, to create scenarios for the most likely earthquakes in southern and northern California [in order to] enable the task force committees to focus on the impacts of the earthquakes. These (eventually seven-plus) scenarios were pioneering efforts to quantify the impacts of the ground motions on structures, lifelines, infrastructure, hospitals, schools, etc. and painted a picture story of what an earthquake like [that of] 1857 or 1906 would be like in today's physical environment. (Rob, personal communication, September 3, 2013)

Structural engineers, urban planners, and seismologists are very active in the adoption of state laws that require California to construct buildings that are more earthquake-resistant, specifically through zoning, housing policy, good governance, and infrastructure policy (ABAG, 2013). Building codes were later reinforced in 1970 and 1990. Until the 1970s, it was the norm that local authorities published their own building codes. The California Building Standards Code (California Code of Regulations, Title 24) was created only in 1978 as an amalgamation and reorganization of existing codes. Following the San Fernando Valley earthquake in 1971, Alquist introduced in 1972 the Alquist-Priolo Earthquake Fault Zoning Act prohibiting "the siting of most structures for human occupancy across traces of active faults that constitute a potential hazard to structures from surface faulting or fault creep"[16] (Alquist-Priolo Earthquake Fault Zoning [AP] Act, 2012). Also in 1972, California governor Ronald Reagan established the Governor's Earthquake Council, which was followed in 1974 by the Seismic Safety Act—the bill that created the Seismic Safety Commission.[17] Jane, a structural engineer, recalled how the concerns over building safety progressively moved from safer schools to safer hospitals to the idea that some buildings welcoming the public should be reinforced so as not to be life-threatening, but rather what she called "life-safe."

[16] He worked on the Alfred E. Alquist Hospital Facilities Seismic Safety Act of 1983.

[17] Now the California Seismic Safety Commission

> [After] the 1971 San Fernando earthquake, the State of California realized that hospitals are facilities that really should be standing up after an earthquake. And they transferred the design of hospitals to an agency called OSHPD—Office of Statewide Health Planning and Development.[18] Every hospital has to go through a plan check and a design check by OSHPD. New hospitals are doing pretty well, but we still have this huge inventory of hospitals designed before. Finally, 1989 happens, the Loma Prieta earthquake, and once again, people are thinking, "Okay, our hospitals are still vulnerable …," and so they started to—this is just the State of California—create a group that looked into a Senate bill. They eventually passed SB1953[19] and came up with deadlines for when you need to evaluate your hospital, when you need to have it "life-safe," and when you need to have it meet full code. And this was 20-odd years after the 1971 earthquake. So, if you look at [it] from start to finish, that's a 50-year process, almost. It's huge. But that came from the top down. And the reason that it takes so long is that it's outrageously expensive, it's billions of dollars. (Jane, personal communication, 2010)

In 1986, the State of California passed the "Unreinforced Masonry Building Law" (SB547), which required each jurisdiction in California's "seismic hazard zones" to reduce unreinforced masonry hazards.[20] For earthquake preparedness within a community, materializing resilience

[18] Office of Statewide Health Planning and Development.

[19] SB 1953 was, in fact, written in 1994 after the Northridge earthquake as the continuation of the Alfred E. Alquist Hospital Seismic Act also known as the "Hospital Act". "The Alfred E. Alquist Hospital Seismic Safety Act ("Hospital Act") was enacted in 1973 in response to the moderate Magnitude 6.6 Sylmar earthquake in 1971 when four major hospital campuses were severely damaged and evacuated. Two hospital buildings collapsed killing forty-seven people. Three others were killed in another hospital that nearly collapsed. In approving the Act, the Legislature noted that: '[H]ospitals, that house patients who have less than the capacity of normally healthy persons to protect themselves, and that must be reasonably capable of providing services to the public after a disaster, shall be designed and constructed to resist, insofar as practical, the forces generated by earthquakes, gravity and winds' (Health and Safety Code Section 129680). When the Hospital Act was passed in 1973, the State anticipated that, based on the regular and timely replacement of aging hospital facilities, the majority of hospital buildings would be in compliance with the Act's standards within 25 years. However, hospital buildings were not, and are not, being replaced at that anticipated rate. In fact, the great majority of the State's urgent- care facilities are now more than 40 years old" (California Seismic Safety Commission, 2001: 3).

[20] These structures are, still today, among the most hazardous buildings in the event of an earthquake.

into generalizable regulation remains a complicated process that involves the designing of seismic provisions into law and the establishing guidelines that will be later be implemented in the field. Since 1933, more than 80 laws have been passed in California to promote earthquake safety and mitigate damage and injury.

6.1.1 Emotional Bonding and Politics: The Co-construction of Resilience

Concerned residents—homeowners, businessmen, inventors, research scientists, government officials, engineers, architects, contractors—mentioned by Tobriner (2006) came together, as did community organizers, local officials, building departments, emergency managing offices, churches, and schools, to determine what a safer Bay Area could look like. In the 1960s Rob had studied at UC Berkeley and been inspired by the work of Saul Alinsky, a Chicago native and community organizer who had written the book Rules for Radicals (Alinsky, 1989).[21] At a moment when the modality of resilience and the infrastructure that will support it needed to be invented, Alinsky's arguments made a difference, as Rob explains:

> We wanted to empower people. We wanted them to feel they [had] the control of the situation. We would go from city to city and talk to the building department people, show them our scripts and propose to talk to their supervisor or to city council. We offered them a "planning partnership" which helped them identify the buildings, infrastructures, and systems at risk and what could be done to improve their organization and their capacity to face a disaster. (Rob, personal communication, October 3, 2013)

Focusing on a very different approach than the ones favored at the time, his team brought together scientists and senior political staff to explain what an earthquake would mean for each institution that they worked

[21] Originally active in the labor movement, Alinsky had involved with in some of Chicago's poorest neighborhoods in the late 1930s and rapidly become an inspiration for activists and politicians alike, such as Barack Obama. Inspired but criticized, Alinsky's followers redefined the objectives and methodologies of community organizations and mobilization.

with, including "a Southern California county (San Bernardino County), a large city (Los Angeles), a small city (Westminster in Orange County), and a corporation (Security Pacific Bank)" (Geschwind, 2001: 217).

As Rob recalled, these meetings included a lot of coffee and a lot of listening, even more so than talking. The team motto was "do not focus on *you should.*" Instead, these meetings provided a safe environment to think about the "unthinkable"—the injuries, deaths, and financial devastation that a large catastrophe could provoke. In each of these meetings, the group members act as a translator for government personnel and for private institutions, "making sense of the language of engineers" (Chakos et al., 2002: 7) with a very strategic perspective:

> For each of these institutions, project staff members worked together with emergency preparedness personnel to develop detailed response plans and contingency plans for a large earthquake. Once these prototype plans had been worked out, they could then be transferred to similar entities across the region. In this manner, earthquake planning increased significantly in Southern California, because individual preparedness offices now had detailed models of emergency plans [...] available to them. (Geschwind, 2001: 217)

Within a few years, the team was involved in shaping earthquake preparedness plans for San Francisco, Berkeley, Oakland, Santa Cruz, Palo Alto, and Santa Barbara. In each of these cities, they focused on what they termed "sweet spots": schools, hospitals, and government buildings. In just a decade, Berkeley residents voted for several measures that collectively allowed more than $390 million from local taxes to be used for municipal safety improvements including seismic retrofitting of city buildings and various school safety programs[22] (Chakos et al., 2002).

[22] In four local elections, Berkeley voters approved over $390 million in local taxes to fund mitigation projects. In addition, the city now rebates one-third of its real estate transfer tax, up to a maximum of $1500, for seismic retrofit—Goldfarb was the swing vote in agreeing in 1991 to increase the transfer tax only if the city council agreed to the rebate. As a result, over 39% of Berkeley's 22,000 single-family residences and over 30% of small multi-family buildings now have improved seismic resistance, at a cost to the city of over $10 million in foregone taxes. The Disaster Council—like the Seismic Safety Commission at the state level—has kept earthquake safety on the agenda of the city council (Chakos et al., 2002).

Some have said this was one of the most important "self-inflicted" increases in local taxes in the history of the United States.

Finally, following the 1989 earthquake, large public and infrastructure works began, which made the potential risks even more visible for residents of the Bay Area. As one respondent commented:

> The concern that the Bay Area's residents have had since the Loma Prieta earthquake with regard to infrastructure is visible. They've seen the signs, gone through the disruption, seen the toll bridges and interchanges on highway bridges retrofitted. Their bills for water [were] increased to pay for upgrades to the water system. They've seen the fares and tolls and their property checks and bills change to pay for the retrofitting of the BART system. They've seen [that] this not only hit them personally, in terms of money, but they've also seen the construction happening. I think there is a larger group of people who understand they have to retrofit their homes than used to be the case. (Jane, personal communication, 2010)

There is no doubt that these actions were made possible by the social and political context in which they occurred. For several decades, Berkeley, and the larger Bay Area, had been known for its strong attachment to a lively democratic tradition. In this "atmosphere in which leaders take responsibility for reducing risk" (Chakos et al., 2002), the dedication of this small team reached out to more communities, and cities started to take concrete, visible actions. Does this mean that the Bay Area is now ready for the next Big One? Things are, of course, more complicated.

6.1.2 *Fragile Infrastructure, Fragile Resilience*

One ricochet effect of the 9/11 catastrophe was the nationwide reorganization of 22 separate disaster-mitigation agencies under the supervision of the Department of Homeland Security (DHS). At the time, these agencies accounted for 180,000 employees and 650 separate computer systems, according to Charles Perrow (2006). "The new department merged agencies that, along with their security roles, had responsibilities for such activities unrelated to terrorism as fisheries, river floods, animal diseases, energy reliability, computer crime, citizenship training, tariffs on imports, drug smuggling, and the reliability of telephone networks" (Perrow, 2006). For many disaster-management experts, this "potpourri of unrelated activities" was not helpful in reducing terrorist threats and "may even have reduced our protection from our two other types of disaster, natural and industrial" (Perrow, 2006). [23]

These political changes impacted the local network of earthquake experts: "It was an overnight change," recalled Rob. Within months, his office library in Oakland was closed down and 20 years' worth of documentation and archives was thrown away. In 2003, Arnold Schwarzenegger took office as governor of California; Rob recalls: "He said to the people of California, 'If you need something, call me!'—which doubled the amount of work." These calls largely disturbed the organization of the group, forcing them to answer many kinds of solicitations, even those not related to earthquake preparedness. The expertise and dedication of the group were no longer considered paramount, and its effectiveness became diluted by the arcane world of administration. As time passed with no reminders of the region's seismic activity, the earthquake problem became a low priority on the political agenda: for more than a decade, the message for local governments was essentially "You're on your own." Experts worried about a total disengagement of state and federal government support, hoping that the actions taken during the 1990s would suffice to limit the effects of major catastrophes, including any large-scale earthquakes (Chakos, 2006).

In addition, and despite all these efforts, research was showing that the building codes, as good as they could be, were often not enforced by local building departments (Burby & May, 1998). After more than a century of building resilient infrastructure, "The greatest challenges to California's building departments continue to be: 1) The lack of public awareness and political support for effective code enforcement; 2) The risks of existing vulnerable buildings in earthquakes; and 3) The need to enhance staff sizes and qualifications to keep up with phenomenal growth." (Turner, 2004: 10) reminded structural engineer Fred Turner, another important figure in the Bay Area's earthquake community.

[23]A situation conducive to the disastrous response to Hurricane Katrina in 2005 (Comfort, 2006; Frickel & Bess, 2007). Perrow noted: "While DHS promulgates an 'all-hazards' approach, Hurricane Katrina in 2005 prompted inquiries that disclosed substantial funds were diverted from programs aimed at natural disasters to those focused on potential terrorist attacks. First responder funds, for example, were cut. Funds for anti-terrorist efforts (improved documentation requirements; watch lists; surveillance of mosques, ports, airports and public buildings; and many of the disturbing provisions of the PATRIOT Act) do not always help with the other hazards" (Perrow, 2006).

Today, after 30 years of being involved in the field of disaster preparedness in the Bay Area, Rob has retired, and the task of providing information and advocacy is pursued by a new generation of experts working hard to maintain the level of awareness necessary to good earthquake preparedness. Since Hurricane Katrina, Bay Area earthquake experts have learned their "ABK" ("All But Katrina"), and the turn of the century has brought renewed interest in the question of earthquake safety. Building on their existing network and surfing the wave of new and potentially game-changing projects such as Earthquake Early Warning systems (Lin, 2017; Strauss & Allen, 2016), they still rely on that strong community to integrate new technological systems and reinforce the strengths of the interactions and dependencies of each urban system. They acknowledge that multiple socio-technical systems will come into play: local governments, schools, water, transportation, churches, private companies, gas, communications, electricity, health and human services, and so on.

Experts' continuous efforts to understand the mechanisms of earthquakes and their consequences for built environments have created and developed further the regulatory apparatuses designed to strengthen building construction codes in California, and to avoid—as much as possible—major damage caused by an earthquake.

> The story of earthquake hazard mitigation in twentieth-century California, then, is the story of the progressive impulse among a small group of California scientists and engineers and the persistence of that impulse even after other reformers had become disenchanted with the state. (Geschwind, 2001)

The recollections and personal trajectory composes a situated perspective on the social and historical conditions that led to the emergence of sociotechnical resilience in the San Francisco Bay Area. The experience of living with the earthquake gives sense to these composite sets of actions, provides a space for the sociotechnical resilience to emerge, and defines the contours of a collective space of risk: a world composed and organized around constantly moving tectonic plates. The exploration of Rob's history also tells us another story: that of the fragility of the network of risk awareness, which can at any time be weakened by the wrong kind of attachments (Latour & Girard Stark, 1999).

Resilience is a syncretic social process, an intended consequence of the strength of the network of risk awareness that supports the sociotechnical infrastructure. Such an examination shows the slow and progressive co-construction of scientific and lay knowledge, building codes, public information, progress in structural engineering, and, later, the more systematic cross-evaluation of independency and path dependency that might have created the conditions for resilience in the Bay Area. By focusing on the sociotechnical infrastructures that make resilience possible, experts and earthquake scientists think about resilience not as a generic set of tools and good practices, but as a situated endeavor whose success and failure cannot be attributed to an individual actant alone but to a chain, a network with its own historiography, its own geographical and epistemological dimensions.

In light of recent work by Peter Hall and Michèle Lamont, I want to point to the importance of the politics of this network and its consequence on the different modalities of resilience. Exploring the concept of social resilience, Hall and Lamont noted that it "depends not only on the feature of society on which individuals draw to enhance their capabilities but also on the capacity of communities to mount collective response to challenges." Moving from the individual scale to the more collective one, social resilience can be defined by "the capacity of groups of people bound together in an organization, class, racial group, community, or nation to sustain and advance their well-being in the face of challenges to it" (Hall & Lamont, 2013: 2). Considering the transformability dimension (Amir & Kant, 2017) of the disaster, there remains the question of the large variability of vulnerability—the cultural, economic, and social repertoires (Hall & Lamont, 2013: 14) that different groups are able to mobilize—within the same space.

The considerable effort in education and outreach conducted by experts such as Rob has relied on the idea that tailored recommendations gleaned from constant field work and from an in-depth knowledge of each community's vulnerability will create better preparedness and, in the long term, better resilience. But more granular research needs to be done to reconnect this in the fields of disaster studies.

6.2 The Instauration of the Earthquake Risk in the Bay Area of San Francisco

As I am finishing this book, the instauration of the earthquake risk continues. In 2014, a new hypothesis emerged: the San Francisco Bay Area might not be facing one, but in fact, several earthquakes in its future. "*Everyone is still thinking about a repeat of the 1906 quake,*" said author David P. Schwartz, referring to the San Francisco Earthquake, which, as previously discussed, had a magnitude of 7.9 had sparked a massive fire, killing 3000 individuals, "*[b]ut what happens if every five years we get a magnitude 6.8 or 7.2? That's not outside the realm of possibility*" (Audi, 2014).

In 2013, after decades of latency, San Francisco City Council had finally signed the long awaited Soft-story Ordinance Bond[24] (Arroyo, 2013; Arroyo & Grady, 2013; BONDS, 2010), and Governor Jerry Brown ordered the Office of Emergency Services to develop an early-warning system. Senator Alex Padilla, who had been the sponsor of the latter bill, became interested in the early-warning question while visiting the California Institute of Technology: "*California is going to have an earthquake early warning system, the question is whether we have one before or after the next big quake*" (AP, 2013).

The future will tell if these initiatives have been an important step toward earthquake preparedness for the municipalities of the Bay Area that have been built during the last 150 years are sitting right above a complex system of active seismic fault lines. Two faults seem to be more at risk than others: the Hayward Fault, which runs through the East Bay, underneath the cities of Richmond, El Cerrito, Berkeley, Oakland, San Leandro, Hayward, Fremont, and San Jose; and along the Pacific, the San Andrea Fault, which runs on and off of the shores of Marin County, San Francisco, and the San Francisco Peninsula. Additionally, because earthquakes can trigger landslides and fires, the multiple areas built upon made, soft ground or in zones prone to wild fires cumulate risks.

[24] Mayor Edwin M. Lee signed the Mandatory Soft Story Ordinance, San Francisco's recently approved Ordinance No. 66-13 which requires the retrofitting of all buildings with the following characteristics: Wood-frame structures; those containing five or more dwelling units; those having three or more stories; and those permitted to be constructed prior to January 1, 1978.

In 2010, 7.88 million people were living in one of 12 Bay Area's counties. Among them, earthquake experts and non-experts alike shared the same "risk."

Despite recurring, alarming predictions, large earthquakes and fires are, in fact, very rare. Not totally forgotten, but not totally present either, their existence—to use Souriau's vocabulary (Latour, 2011)—seems incomplete or partial, that is, until they become destructive and—way—too real. Several decades of labor in various scientific fields—collecting data, advocating, training, crafting ordinances, bonds and codes, and evaluating impacts and consequences—have given more substance to earthquake risk's existence. The earthquake as a phenomenon, in the Souriau sense, transforms our understanding of space, the way we think about security and safety, the certitude that we have in science and the understanding of expertise. But like a work of art—which was the example given by Souriau—the risk of an earthquake, can fail to exist; in this case, the failure is not only the failure of definition, but moreover, a failure of instauration.

Earthquake risk in the Bay Area is collectively instaured by risk-aware residents and experts who called themselves the Earthquake Junkies. Describing these moments of instauration, I have shown how a network of attention to earthquake risk has been built along different lines, allowing for the coexistence between expertise and common knowledge, science and experience. In the sense this instauration falls within this context of a redefinition of the concept of expertise.

It is through the personal, intimate, experiences of living with the earthquake that experts-residents or experts-amateurs finally give a moral perspective on what they describe as an "addictive" attachment to the acts of risk prevention. In the process, the question of knowledge has been raised many times. This is not only because earthquake science is still in the process of emerging—what science is not?—but also because of the division between the knowledge of what has been considered valid and invalid. Crossing together concepts and methodologies used by seismologist to measure the perception and theoretical reflection, which draw a continuum between James pragmatism and ANT, we have deployed another example of the making of a scientific though: the movement which transforms experts into amateurs. Finally, what has emerged from this field research is a figure of the expertise and the earthquake scientist who is borrowing for the portrait of the expert-amateur defined by researchers in ANT, becoming an activist, a stockperson, engaged in an ongoing dialogue with the communities he is serving.

Reflecting on Mialet's description of Stephen Hawking collective practices (Mialet, 2012), Michael Lynch recently stated that previous understandings of expertise have been staying "*far too distant from (...) networks and singular circumstances*" and have therefore "*stated intention to treat expertise as real is a general position statement rather than a demonstrable finding.*" As oppose to these static definitions, he argued that Mialet's allows us to see that:

> The identity and authority of an expert is not a consequence of verbal attributions made in isolation. Instead, it is a continual project, involving many participants working in organized, embodied, and intimate ways within specific institutions. (Lynch, 2014)

"Waiting for the Big One" is paying attention to the creation of the idea of risk. So is the earthquake instauration. Earthquakes produce movement: the movement generated by the tectonic plates, of course, the movement produce by the different instauration of the earthquake phenomenon, and finally the movement provoked by the mobilization around an organized project of earthquake preparedness. Following James's observations after the 1906 Earthquake, I have shown how these multiple dimensions cannot be reduced. Rather, each influence another, creating a unique form of relationships and knowledge that together encapsulate the ways in which people "wait" for the next Big One.

I have shown that while instauring the risk of earthquakes, Earthquake Junkies initiate a collective, common world of shared existence and experience. It is the experience of living with the earthquake—of waiting for it, fearing it, remembering it, and getting ready for the next one—that allow different earthquake ontologies to emerge, resulting in making the definition of the earthquake risk and expertise, and finally the instauration of a collective space of risk—a "we" which is not the community that Saul Alinsky dreamt about, but something else—a world composed of beings and experiences.

Reconsidering the articulation between expertise and laity, between knowledge and "non knowledge," has been one way to question "the economy of risk," to use a Boltanski and Thevenot's expression (Boltanski & Thevenot, 2006). These reconfigurations of the moral and pragmatic figure of the expert and practice of expertise are calling for more investigation when confronted to the figure of the public as define by revisited by contemporary scholars.

> The public is conceived as a collective instance potentially constituted by those who are affected by problems, suffer the indirect consequences of interaction and are interested to master them. The defined exploration of the "consequences," for further treatment of the problems, soon constitute the operator of the political experience through which a community oriented toward its own perfection ideally get determined. In the way, trough the model of the scientific inquiry, it requires, up-front, the training of the scientific capacity of all as much as each can participate, under the form of the public, to the realization of an ideal of regulation and conformation of the community.[25] (Stavo-Debauge, 2004)

The Earthquake Junkies of the Bay Area are not an isolated tribe with bizarre practices, they are not a peculiar public. As I was discussing these questions during the 2013 4S Conference in San Diego, Scott Frickel, then Associate Professor at Washington State University, who himself had to relocate to Seattle after Hurricane Katrina ravaged New Orleans, and whose work on expertise has become seminal in the field (Frickel & Bess, 2007; Frickel, Campanella, & Besse, 2009), remarked that many—if not most—researchers involved in the study of risks and disasters have a deep emotional and intellectual connection with their subject of research. I could not agree with him more. Genuinely bricoleurs, full time residents, dedicated scientists, concerned citizens, empathic researchers, the Earthquake Junkies are part of a larger community, navigating between scientific conferences and city halls meetings, who is reshaping the definition of knowledge by reconnecting scientific knowledge with experience.

[25] "*Le public y est conçu comme une instance collective potentiellement constituée par ceux qui sont affectés par des problèmes,subissent les conséquences indirectes d'interactions et sont intéressés à leur maîtrise. L'exploration réglée des« conséquences», en vue d'un traitement des problèmes, constitue dès lors l'opérateur de l'expérience politique par laquelle se détermine idéalement une communauté orientée vers sa propre perfection. En eminent à travers le module de enqueue identified, Elle exile, en amont, la formation de capacités spécifiques de tous en tant que chacun participe, sous la forme du public, à la réalisation d'un idéal de régulation et de conformation de la communauté*" (Stavo-Debauge, 2004).

This investigation of the earthquake risk in the San Francisco Bay Area has been an attempt to sew together objects, thoughts, and theories that are all part of the instauration of earthquake risk. In this context, the goal of this work has been to capture the entanglement of the relationships to risk from a perspective which has tried to be neither moralist nor normative. A second objective was to find a way to describe, in theoretical terms, what I observed in my field research. In this sense, if the organization of this work presents the theoretical part first, it is done with the purpose of following the academic model of research, but it does not fully inform the multiple movements back and forth between books and the transcripts of my interviews.

This work has been empirical, in a Jamesian sense, observing the density of experiences through the relation, the distance, between actants. Following the programmatic research agenda defined as "pragmatic sensibilities", I have followed:

> a certain interest for the exploration of a world in transformation, sharing the desire to describe it more finely et to find for that the tools and the concepts susceptible to take seriously what change and what provoke change.[26] (Cantelli, Pattaroni, Roca, & Stavo-Debauge, 2009)

The condition in which the field research is conducted does have consequences on the obtained results. As a resident of the Bay Area since 2008, I have had the time to be shaped by the risk of a local earthquake while simultaneously trying to define it. Revisiting these last years in my mind, it has become easier for me to identify the transformations of the understanding of living with this risk at different moments. From totally skeptical to very concerned, from very detached to extremely involved, I have experienced the different stages of attention to risk. Living in the Bay Area and conducting research on risk presented the opportunity to experience this intermediate layer of knowledge personally. Sharing the everyday-ness of the potentiality of a disaster, being attentive to the details of the relationships between the different dimensions of earthquake risk, and the transformation of mundane objects

[26] "*D'une certaine manière, nous voulons défendre l'idée que les sensibilités pragmatiques se rejoignent avant tout dans un certain intérêt pour l'exploration d'un monde en transformation, partageant une envie de le décrire plus finement et de trouver pour cela les outils et les concepts susceptibles de prendre au sérieux ce qui change et ce qui fait changer*" (Cantelli et al., 2009).

of the environment, have taken the form of a "questioning situation" (Stengers, 2009.)

This proximity between my own experiences and the subject of my research forced me to reopen the discussion of common assumptions in the scientific world about the ignorance, foolishness, or irrationality of residents living in risk-prone areas. In the Bay Area, the people with whom I interacted every day—the ones who I interviewed, my friends or my colleagues at the University of California, Berkeley—were not at all ignorant, foolish, or irrational. But still, they were living in a high and dangerous seismic zone. This apparent contradiction was somehow reinforced in a context where the definition of risk was challenged by the experts themselves, where my naïveté was sometimes mocked with crude irony during my interviews, where the diversity of experiences and forms of knowledge mobilized by the earthquake community did not seems to present any homogeneous perspective.

I have tried to build a theoretical framework in which the different epistemologies and ontologies of the people whom I have met, and who have given me their valuable time, could be engaged in an active discussion. It is therefore symptomatic that the main inspiration of this work comes from a field that specializes in mapping controversies. Coming for the field of Sciences and Technologies Studies and Geography, informed by Anthropology, the second chapter provided the necessary theoretical framework to develop my project. In this chapter, I showed how, in order to understand the relationship of risk as defined both as an event, a cause, a probability or a statistical expectation and the ways in which decisions are made under the previously cited conditions, it needs to be considered as an actant, in the ANT understanding of the term.[27]

[27] "In non-technical contexts, the word "risk" refers, often rather vaguely, to situations in which it is possible but not certain that some undesirable event will occur. In technical contexts, the word has several more specialized uses and meanings.

Five of these are particularly important since they are widely used across disciplines:

1. Risk: can unwanted event which may or may not occur.
2. Risk: the cause of an unwanted event which may or may not occur.
3. Risk: the probability of an unwanted event which may or may not occur.
4. Risk: the statistical expectation value of an unwanted event which may or may not occur.
5. Risk: the fact that a decision is made under conditions of known probabilities ('decision under risk' as opposed to 'decision under uncertainty')" (Hansson, 2012).

Examining the how definitions of the concepts of risk and disaster have been mobilized, I have identified continuum of definitions, which from the most determinist to the most constructivists, have organized the relative position of the actants at play.

I have later argued for an understanding of risk and disaster that offered the possibility of developing a contemporary, pragmatic approach to the concept of risk, following the path opened up by James when he visited San Francisco after the 1906 Earthquake. As a reminder of this approach, I have used James's quotations as a thread to walk me through the different aspects of the earthquake, which I needed to consider together:

> For "science," when the tensions in the earth's crust reach the breaking point, and strata fall into an altered equilibrium, earthquake is simply the collective name of all the cracks and shakings and disturbances that happen. They are the earthquakes. But for me the earthquake was the cause of the disturbances, and the perception of it as a living agent was irresistible. (James, 1906)

As much as Hansson's definition of risk captures the different "mode of existence" of risk—"an unwanted event which may or may not occur, [...] the cause of an unwanted event which may or may not occur, [...] the probability of an unwanted event which may or may not occur, [...] the statistical expectation value of an unwanted event which may or may not occur"—my field research has demonstrated that the "question of known probability" and the "decision under uncertainty" (Hansson, 2012) were not distinct and opposed regimens of knowledge, but rather, it could be described as a continuum that included different forms of legitimate, but also contested knowledge. To trace this continuum, I have used the concept of "instauration," developed by the French philosopher Etienne Souriau:

> Instauration and construction are clearly synonyms. But instauration has the distinct advantage of *not* dragging along all the metaphorical baggage of constructivism — which would in any case be an easy and almost automatic association given that an artwork is so obviously 'constructed' by the artist. To speak of 'instauration' is to prepare the mind to engage with the question of modality in quite the opposite way from constructivism. To say, for example, that a fact is 'constructed' is inevitably (and they paid me good money to know this) to designate the knowing subject as the

> origin of the vector, as in the image of God the potter. But the opposite move, of saying of a work of art that it results from an instauration, is to get oneself ready to see the potter as the one who welcomes, gathers, prepares, explores, and invents the form of the work, just as one discovers or 'invents' a treasure. (Latour, 2011: 10)[28]

The field research made clear that earthquake scientists and geographers, as researchers coming from the empirical sciences who are forced to deal with the materiality of their object of study, have long taken perceptions and emotions seriously, not as a by-product that needed to be "cleaned up," but as legitimate actants in themselves that need to be given full attention. The work of the British geographer Nigel Thrift and non-representational theory, allowed me to find a path to reconnect to some of James's pragmatic insights with my geographical, spatial thinking anchored in the geography of risk.

Starting with the hypothesis that earthquake risk is a complex, and not always visible object, the third chapter questioned the visibility of this risk. In search for traces of earthquakes in the landscape, through the memories of long-time residents and via commemorations, I have argued that earthquakes have had a strong influence on shaping the space of the San Francisco Bay Area, but that this influence often remains invisible.

In this chapter I challenged the assumption that risk is "*the fact that a decision is made under conditions of known probabilities*" (Hansson, 2012), showing how the controversies about the space of risk in the Bay Area have tended to erase the traces of risk from the landscape. Focusing on discontinued dimension of the risk I shown that the earthquake is not easily visible: made invisible in the early twentieth century to avoid having investors be turned away, the contradictory definitions of risk and disaster in the last decades have made difficult the possibility of commemorations and remembrances.

Observing the invisible transformations that cohabitation with the risk of earthquakes creates, this chapter is at the cross section between territoriality, science, and action. This place in not coincidental; in fact, it was probably one of the most important results from my fieldwork. I discovered the transformative experiences of past and distant events that

[28] "The French legal term for someone who discovers a treasure is actually the 'inventor'… French is constructivist by construction!" (Latour, 2011: 10).

Earthquake Junkies have that keep them so deeply rooted and attached to their subject of research.

The hypothesis of the fourth chapter focused on the experiences of living with earthquake risk, and its capacity to transform the categories of knowledge, experience, and subjectivity. Here, I argued that the "quality" of the idea of an earthquake—as a potential concern or a threat—can only be found while paying attention to the density of experiences and in the everyday coexistence with active seismic faults. Looking at the different existences of the earthquake phenomena allows us to approach the elusive quality of this risk, but also its performative dimensions. In this chapter, I have tried to understand how, in the absence of the Big One how does the earthquake exist for residents. Indeed, if the Big One remains on the horizon of the possible, earthquakes still come in all shapes in the Bay Area contributing to define in a kaleidoscopical way the multiple existences of the earthquake.

Building on Souriau's concept of instauration, I have focused mainly on the immaterial dimension of the earthquake existence, questioning how the earthquake as a phenomenon is able to transform the world as we know it. The deployment of the immaterial dimensions of the earthquake led me to consider the unexplored field of emotions and perceptions that are connected to the idea of an earthquake. Thus, in the Latourian definition of the mode of existence, this third chapter is about habit, or [HAB]:

> *Habit is the technical term that has been chosen to designate a mode of existence* [HAB], *which is characterized by the designation of a movement directed towards the course of action and therefore away from the preposition. Philosophers of habit, few in number because it has been seen as a doxa, and a form of ignorance, due to its implicit nature, have always insisted on the fact that courses of action require additional adjustment in order to be carried out.* (Latour, 2012, 2013)

Convocating what is often considered a too messy material for research, I have examined the moment when knowledge get hybridized by the mundane experience of habit, as these actants present a "radical alterity" (Latour, 2014). The first part of this chapter focused at other forms of tremors, like the small earthquakes that people feel far more frequently than larger ones. Because they "touch" residents, these events trigger actions and reflections about the Big One.

I took seriously a certain particularity of San Francisco Bay Area, namely, the capacity to make fun of earthquakes, as well as the gentle chiding of people fear, which are part of the everyday, earthquake mundane existence. As much as the memory of the Oakland Fire was hidden in the museum archives, the awareness of the earthquake, the deep understanding that the Bay Area could be destroyed in couple of seconds, lay hidden inside the ability to joke about these events. Humor acts as distanciation; it translates the interwoven processes of close or distant attachment, characteristic of Bay Area residents' ability to "deal" with earthquake risks. Distant earthquakes, like the ones in New Zealand, Haiti, or Japan bring the possibility of earthquakes terribly close, while jokes gently keep the real possibilities and realities at bay.

Looking at the earthquake as a mediated event, this chapter questioned the usual distinction between science—"good information"—and lay knowledge content—"bad information." Researchers in the field of Media Studies have long noted that the internet is a space of contested knowledge (Abe, 2013; Seidman, 2008). In the context of this research, it was interesting to see how citizen-science and blogs have played very important roles in informing the public about the multiple dimensions of potential catastrophes, while incorrect—or partial—information was relayed by traditional media.

As echoed from the previous chapter, which emphasized the modification of space during and after a disaster, the third chapter addressed the transformation of residents when confronted by the potentiality of the destruction of their familiar environments, or worse, the deaths of family and friends. Some interviewees have recognized the highly emotional impact that past disasters have had on their psyches: how deeply structuring those experiences of death and destruction have been in their lives—whether because they decided to engaged fully in avoiding such events, or because they fundamentally have changed their lives so as to avoid dealing with this form of major destruction ever again.

My hypothesis questioned the ways in which this hybridized knowledge creates forms of engaged expertise, which has had the effect of constantly reshaping the urban space of the Bay Area, as well as the subjectivity of experts in the field. For decades now, earthquake scenarios have defined the scientific existence of earthquakes. In this chapter, I have decomposed at the ways in which scientific communities have framed this existence. Doing so, I have deployed part of the very dense network of activity, which crystallizes in the earthquake map and hybridises the definition of science and expertise.

In the last chapter, I have explored the scientific apparatus, which has been mobilized in the last few decades to better define earthquake risk. I have shown how the definition of risk is a collective endeavor, which requires the energy and dedication of many scientists and experts. Challenging the argument that earthquake scientists look at the world from an ivory tower, this chapter has shown how experts engage with their object of study, tied to ethical and social considerations.

The earthquake-risk scenario focuses on calculations, whether they concern the probability of a fault rupture, or the insured or uninsured costs incurred for a particular rupture in a given place. Scenarios use data constructed according to the methods defined and accepted as scientific process (Daston, 1991). They brought together the actants in order to estimate their interactions and their potential dangers—the strengths or weaknesses of the soil, buildings, and infrastructure. To obtain their objective, these scenarios focused their attention on the interactions of tectonic-plate movement (Modified Mercalli Intensity, magnitude, liquefaction); consequences of plate movements (fire-related damage, floods, landslides); potentially aggravating factors (wind conditions and other adverse meteorological conditions); and on buildings (retrofitted, not retrofitted, soft-story, unreinforced, masonry); public facilities (schools, hospitals, state and federal buildings); infrastructure (water, sewer, gas, transportation, bridges, piers, tunnels); population (prepared or not, injured, dead, displaced, or traumatized)[29]; the economic situation (sales, taxes, revenue, insurance, mortgage defaults),[30] all just to name a few. And this does not take into consideration some of the cumulative effects of some of these aspects occurring together at particular moment in a particular event.

In the final chapter, we have explore how scientific knowledge about earthquakes is embedded in a larger set of political preoccupations about the safety of Bay Area residents. Successes and failures of science cannot be evaluated without looking at the work—often invisible—done by community mobilizers. In the case of the Parkfield Experiment, for

[29]The population of the Bay Area is estimated to be 7.4 million residents in 2.7 million households.

[30]Projections of those employed are estimated to be 3.4 million with jobs. The regional economy has been, in 2009, estimated to be approximately $300 billion.

instance, residents were sensitized to the risk of earthquakes, and new research hypotheses have emerged from the ones which were invalidated. Taking all of this a step further, these pieces of information were later translated into an apparatus of regulations and laws specifically designed to avoid damages during the next major earthquake.

In this work I have examined the necessary operation of translation between the tools used to define and measure an earthquake and the individuals who are living in risk zones. Describing these varied aspects of the instauration of earthquake risk, it became visible that this instauration is in fact a co-construction of both the risk and the expertise—the scientific knowledge and the experience of living with the earthquake. Considering the different dimensions of experience, this work has looked at the invisible actants—"the traces" which both define epistemic territories, cultural practices and more broadly the space of risk and disaster, at time hidden and then made visible again.

The instauration of the earthquake continues to transform the individual lives of concerned people. In a sense, the career evolution of some of the individuals whom interviewed for this research tended to prove that scientists and experts move along a network in which they maintain a strong connection both with their experiences as residents, and with the local, state, and federal institutions in charge of managing seismic risks.

Since the time of my interview, J.15, who at the time was so engaged in improving seismic resistance in her school district, joined the California Earthquake Authority, an organization founded in 1996 by the California Legislature to develop earthquake insurance policies. Another interviewee, T.4, after working for 45 years in earthquake preparedness helping California Legislators draft earthquake-preparedness bills, has since become the President of the Earthquake Engineering Research Institute, and works as senior consultant for GeoHazards International, an NGO involved in seismic-risk mitigation around the world.

The Northern California Chapter of the Earthquake Engineering Research Institute has presented a special award to my interviewee, J.8, for "Lifetime Achievement in Earthquake Risk Reduction," along with a "certificate of appreciation" from the Association of Bay Area Governments and a proclamation from Oakland Mayor Jean Quan. In addition, A.28, a community organizer for more than three decades, has been working with both the city of Palo Alto and U.C. Berkeley to assess resilience in the case of major earthquakes. D.21, who has been a member of San Francisco's Community Action Plan for Seismic Safety and part of the team developing the FEMA P-807 Guidelines (Guidelines for Seismic Retrofit of Weak-Story Wood-Frame Buildings), is now working for the city of San Francisco on their "soft-story" retrofit program. M.3, who participated in the U.C. Berkeley earthquake preparedness project, and who has written widely on disaster recovery, has received the Green Star Award from the United Nations for her work in the post-disaster reconstruction projects in China and Haiti, and in 2013, received the U.C. Berkeley Chancellor's Award for Public Service for Research in the Public Interest. R.8 has pursued his vision of an early-warning system, and gained state and national recognition when he helped pass Senate Bill No. 135, which states:

> The Office of Emergency Services, in collaboration with the California Institute of Technology (Caltech), the California Geological Survey, the University of California, the United States Geological Survey, the Alfred E. Alquist Seismic Safety Commission, and other stakeholders, shall develop a comprehensive statewide earthquake early warning system in California through a public-private partnership. (Padilla, 2014)

Of course, many others of the persons I have interviewed have continue to active as well. A.M. 20, pursued her work on emergency preparedness with the California Public Health Systems Research organization and the California Emergency Services Association, is giving lectures and being as part of a community advocacy team to make emergency preparedness accessible to non-profits. R. 28, despite being "really retired," as his LinkedIn page indicates, continues to be passionate about training the next generation of Earthquake Junkies. And R.5 has become a writer; she has written about change and transformation in a 2014 piece entitled "The Next Big Step":

> It's been the story of my life: unexpected deviations from The Plan, unintended consequences, and things that seem to work themselves out even when the best laid plans have gone astray. [...] We moved into this house as a family of four in 1984. By 1986, we were a family of five. And in 1991, the place burned to the ground. We rebuilt in the same spot, and as project manager of the rebuilding I had a hand in every element that went into the new house. Now, over twenty years later, our three kids are grown up and married with families of their own. The house that once felt so big can no longer accommodate our growing crowd on those too-rare

> occasions when we all get together. Just thinking about packing up and moving leaves me with feelings of dread (and denial at the inevitability of it), but a big change also carries with it a certain air of excitement. I'm just not sure I'm ready to make that change quite yet. (Nye, 2014)

In-between earthquakes, life goes on: kids have grown up living with earthquake risks their whole lives. This is also instauring the earthquake. Since I live in this environment too, and have experienced the joy of having a family of my own while working on this research, everything that I have learned made me reflect on what exactly the risk consists of, who are the experts, and how to create good risk-prevention policies, all while packing the disaster kits that my kids' schools ask for each year: a stuffed animal, a family picture, a bottle of water, two cereals bars, and a red flashlight.

I might have become an Earthquake Junkie.

References

ABAG. (2013). *Resilience initiative—Building a disaster resilient Bay Area*. Retrieved from http://quake.abag.ca.gov/projects/resilience_initiative/.

Abe, Y. (2013). Why safecast matters: A case study in collective risk assessment. A STS forum on Fukushima. Berkeley. Retrieved from http://fukushimaforum.wordpress.com.

Adger, N. (2000). Social and ecological resilience: Are they related? *Progress in Human Geography, 24*(3), 347–364. https://doi.org/10.1191/030913200701540465.

Alinsky, S. (1989). *Rules for radicals*. New York: Vintage; a edition.

Alquist-Priolo Earthquake Fault Zoning (AP) Act. (2012). Retrieved from http://www.conservation.ca.gov/cgs/rghm/ap/Pages/index.aspx.

Amir, S., & Kant, V. (2017). Sociotechnical resilience: A preliminary concept. *Risk Analysis, 0*(0). https://doi.org/10.1111/risa.12816.

AP. (2013). Gov. Brown signs bill to develop California earthquake early warning system. *The Washington Post*. Retrieved from http://articles.washingtonpost.com/2013-09-24/national/42352886_1_warning-system-earthquake-u-s-geological-survey.

Arroyo, N. (2013). *No one wants to go first: S.F.'s retrofit timeline*. Retrieved from http://sfpublicpress.org/news/2012-12/no-one-wants-to-go-first-sfs-retrofit-timeline.

Arroyo, N., & Grady, B. (2013). *Earthquake retrofit delays leave thousands at risk*. Retrieved from http://sfpublicpress.org/news/2012-12/earthquake-retrofit-delays-leave-thousands-at-risk.

Audi, T. (2014, May 19). Several earthquakes are bigger threat to San Francisco Bay Area. Multiple major temblors, not a single huge one, pose a bigger risk, study says. *Wall Street Journal*. Retrieved from http://online.wsj.com/news/articles/SB10001424052702304422704579571972490977390.

Boltanski, L., & Thevenot, L. (2006). *On justification: Economy of worth*. Princeton: Princeton University Press.

Bonanno, G. (2004). Loss, trauma, and human resilience: Have we underestimated the human capacity to thrive after extremely aversive events? *American Psychologist, 59*(1), 20–28.

BONDS. (2010). *Proposition A: Earthquake retrofit bond*. San Francisco County, CA.

Burby, R., & May, P. (1998). *Making building codes an effective tool for earthquake hazard mitigation* (College of Urban and Public Affairs (CUPA)—Working Paper, 1991–2000, Paper 3). Retrieved from http://scholarworks.uno.edu/cgi/viewcontent.cgi?article=1002&context=cupa_wp.

California Seismic Safety Commission. (2001). *Findings and recommendations on hospital seismic safety*. Sacramento: State of California Seismic Safety Commission.

Cantelli, F., Pattaroni, L., Roca, M., & Stavo-Debauge, J. (2009). Sensibilités pragmatiques. Enquêter sur l'action publique—Introduction. In *Sensibilités pragmatiques. Enquêter sur l'action publique* (p. 26). Bruxelles: Peter Lang.

Chakos, A. (2006). *Message to locals in California disasters: YOYO-You're on your own*.

Chakos, A., Schulz, P., & Tobin, L. T. (2002). Making it work in Berkeley: Investing in community sustainability. *Natural Hazard Review, 3*, 55–67.

Comfort, L. K. (2006). City at risk: Hurricane Katrina and the drowning of New Orleans. *Urban Affairs Review, XX*(X), 1–16. https://doi.org/10.1177/1078087405284881.

Comfort, L. K., Boin, A., & Demchak, Chris, C. (2010). *Designing resilience: Preparing for extreme events* (1st ed.). Pittsburgh: University of Pittsburgh Press.

Daston, L. (1991). Marvelous facts and miraculous evidence in early modern Europe. *Critical Inquiry, 18*(1), 93–124. Retrieved from http://www.jstor.org/stable/1343717.

Folke, C. (2006). Resilience: The emergence of a perspective for social-ecological systems analyses. *Global Environmental Change, 16*(3), 253–267. https://doi.org/10.1016/j.gloenvcha.2006.04.002.

Frickel, S., & Bess, V. (2007). Katrina, contamination, and the unintended organization of ignorance. *Technology in Society, 29*, 181–188.

Frickel, S., Campanella, R., & Besse, V. (2009). Mapping knowledge investments in the aftermath of Hurricane Katrina: A new approach for assessing regulatory agency responses to environmental disaster. *Environmental Science and Policy, 12*(2), 119–133.

Geschwind, C.-H. (2001). *California earthquake, science, risk and the politic of hazard mitigation*. Baltimore and London: The Johns Hopkins University Press.

Guarnieri, F. (2017). The Fukushima Daiichi nuclear accident: Entering into resilience faced with an extreme situation. In J. Ahn, F. Guarnieri, & K. Furuta (Eds.), *Resilience: A new paradigm of nuclear safety: From accident mitigation to resilient society facing extreme situations* (pp. 1–17). Heidelberg and New York: Springer. https://doi.org/10.1007/978-3-319-58768-4.

Hall, P. A., & Lamont, M. (2013). Introduction. In P. A. Hall, & M. Lamont (Eds.), *Social resilience in the neoliberal era* (pp. 1–34). New York: Cambridge University Press.

Hansson, S. O. (2012). Risk. In S. O. Hansson (Ed.), *The Stanford encyclopedia of philosophy*. Stanford: Stanford University. https://plato.stanford.edu/entries/risk/.

Inventory of the Joint Committee on Seismic Safety Records. (n.d.). Retrieved January 12, 2017, from http://www.oac.cdlib.org/findaid/ark:/13030/tf0290005q/.

James, W. (1906). *On some mental effects of the earthquake. The Youth's Companion*. Reprinted in James, H. Jr, 1911, *Memories and Studies* (Henry James, Jr.). Longmans, Green, & Co. (Vol. June). New York.

Jasanoff, S. (2014). Genealogies of STS. *Social Studies of Science, 42*(3), 435–441. https://doi.org/10.1177/0306312712440174.

Latour, B. (2011). Reflections on Etienne Souriau's Les differents modes d'existence. In G. Harman, L. Bryant, & N. Srnicek (Eds.), *The speculative turn continental materialism and realism* (Anamnesis, pp. 304–334). Melbourne: re.press. Retrieved from http://books.google.com/books?id=coBDqJQeAQYC&pg=PA304&lpg=PA304&dq=Latour+Souriau&source=bl&ots=wHqRXjOttu&sig=GjPBBQjrUsrgrRgvWeykV9m-6Q7Q&hl=en&sa=X&ei=XDFnUe_ZNoXmiwK7toDADQ&ved=0CGIQ6AEwAw#v=onepage&q=LatourSouriau&f=false.

Latour, B. (2012). *Enquête sur les modes d'existence. Une anthropologie des Modernes*. Paris: La Découverte.

Latour, B. (2013). habit [hab], too, is a mode of existence in *An inquiry into modes of existence*. Cambridge, MA: Harvard University Press, p. 265.

Latour. (2014). *Habit*. [hab]. Retrieved May 4, 2019, from http://modesofexistence.org/crossings//#/en/hab.

Latour, B., & Girard Stark, M. (1999). Factures/fractures: From the concept of network to the concept of attachment. *RES: Anthropology and Aesthetics, 36*(Autumn), 20–31. Retrieved from http://www.bruno-latour.fr/sites/default/files/downloads/76-FAKTURA-GB.pdf.

Lin, R.-G. I. (2017, May 3). Earthquake early warning system nets $10.2 million in Congress' budget deal. *Los Angeles Time*. Retrieved from http://www.latimes.com/local/lanow/la-me-ln-earthquake-early-warning-20170502-story.html.

Lynch, M. (2014). Genius, Hawking, and expertise. *Social Studies of Science, 44,* 793.

Mazel-Cabasse, C. (2017). Hybrid disasters—Hybrid knowledge. In J. Ahn, F. Guarnieri, & K. Furuta (Eds.), *Resilience: A new paradigm of nuclear safety: From accident mitigation to resilient society facing extreme situations* (pp. 337–351). Heidelberg and New York: Springer. https://doi.org/10.1007/978-3-319-58768-4.

Mialet, H. (2012). *Hawking incorporated: Stephen Hawking and the anthropology of the knowing subject.* Chicago and London: University of Chicago Press.

Nye, R. (2014). *The next big step.* Retrieved from http://www.risanye.com/.

Padilla, A. (2014). *SB-135 earthquake early warning system.* Senate.

Perrow, C. (2006). The disaster after 9/11: The department of homeland security and the intelligence reorganization. *Homeland Security Affairs, II*(Article 3). Retrieved from https://www.hsaj.org/articles/174%0A%0A%0A%0A.

Reghezza-Zitt, M., & Rufat, S. (2015). *Resilience imperative: Uncertainty, risks and disasters.* London and Oxford: ISTE Press and Elsevier.

Reghezza-Zitt, M., Rufat, S., & Djament-Tran, G. (2012). *What resilience is not: Uses and abuses.* Retrieved from http://cybergeo.revues.org/25554.

Seidman, S. (2008). *Contested knowledge: Social theory today* (4th ed.). Oxford: Wiley-Blackwell.

Stavo-Debauge, J. (2004). Le pragmatisme et son public à l'épreuve du terrain. Penser avec Dewey contreDewey. In L. Quere & B. Karensti (Eds.), *Raisons Pratiques, 15* (Editions d, Vol. 15). Paris: Editions de l'EHESS.

Steinbrugge, K. (1968). *Earthquake hazard in the San Francisco Bay Area: A Continuing problem in public policy.* Berkeley: Institute of Governmental Studies, University of California.

Stengers, I. (2009). *Au temps des catastrophes; résister a la barbarie qui vient* (les empêch). Paris: La Découverte.

Strauss, J. A., & Allen, R. M. (2016). Benefits and costs of earthquake early warning. *Seismological Research Letters, 87*(3), 765–772. https://doi.org/10.1785/0220150149.

Timmerman, P. (1981). Vulnerability, resilience and the collapse of society: A review of models and possible climatic applications. *Environmental Monograph.* https://doi.org/10.1002/joc.3370010412.

Tobriner, S. (2006). *Bracing for disaster: Earthquake-resistant architecture and engineering in San Francisco, 1838–1933.* Berkeley: The Bancroft Library, University of California Berkeley.

Turner, F. (2004). Seventy years of the riley act and its effect on California's building stock. In *13th World Conference on Earthquake Engineering*. Vancouver.

Weichselgartner, J., & Kelman, I. (2015). Geographies of resilience: Challenges and opportunities of a descriptive concept. *Progress in Human Geography, 39*(3), 249–267. https://doi.org/10.1177/0309132513518834.

Wikipedia Contributors. (2018). Unreinforced masonry building. In *Wikipedia, The Free Encyclopedia*. Retrieved from http://en.wikipedia.org/wiki/Unreinforced_masonry_building.

Afterword

In recent years, researchers have pointed out that what is often described as citizen science encompasses distinctive modes, often understood as differentiated sets of practices, purposes, and objectives (Eitzel et al., 2017; Kullenberg & Kasperowski, 2016; Selin et al., 2016; Traweek, 2013). Yet others, working on the role of data (and data science practices), have contributed to a better understanding of the processes by which heterogeneous data are integrated, allowing for the emergence of interdisciplinary practices (Boix Mansilla, Lamont, & Sato, 2016; Borgman, 2013; Jirotka, Lee, & Olson, 2013; Leonelli, Rappert, & Davies, 2017). This proliferation of digital tools, objects, and practices has led to what Marres has described as a "redistribution of research" and a "redistribution of methods" that recognizes the contributions of various agents, "researchers, research subjects, funders, providers of research materials, infrastructure builders, interested amateurs, and so on," (Marres, 2012: 140) and the modalities of enactments that are often hard to pin down. As Star and Griesemer noted almost three decades ago, "[M]ost scientific work is conducted by extremely diverse groups of actors—researchers from various disciplines, amateurs and professionals, humans and animals, functionaries and visionaries. Simply put, scientific work is heterogeneous" (Star & Griesemer, 1989: 391–392). Researchers have also noted that these data practices and modes of producing knowledge emerge from organizational settings, standards, and norms that define collaboration and interdisciplinarity in scientific arenas (Aronova, 2017; Landström, Whatmore, and Lane, 2011; Riesch

C. Mazel-Cabasse, *Waiting for the Big One*,
https://doi.org/10.1007/978-3-030-15289-5

and Nowotny, 2017), as well as by the virtues and political consequences attributed to what has often been thought of as—and criticized for being—a one-dimensional relationship between experts and non-experts (Allen, 2011; Kimura and Kinchy, 2016; Lidskog, 2008; Lynch, 2014; Wynne, 1996).

Following what has been referred to as the ontological turn in science and technology studies (Law and Lien, 2013; Lynch, 2013; Mol, 2013), researchers have acknowledged that there may be different ways to understand ontological questions that are "in actuality decided through specific, historical, cultural, technological, scientific interventions" (Marres, 2013: 423). Expanding our understanding with regard to how citizen science is performed opens up the possibility of reflecting on the specific types of assemblages from which these modes emerge and their specific existences. It is also an invitation to reflect on the integration processes (Akera & Mohsin, 2016; Gerson, 2012; Star & Griesemer, 1989), the communities (Knorr-Cetina, 1999), and the imaginations (Jasanoff and Kim, 2015) that "make" science. Questioning the modality of citizen science therefore requires thinking about the ways in which science is conducted: the *modus operandi* (Whatmore, 2009) according to which the order of things is determined. It also requires a description of how the work of sense-making and knowledge-building is accomplished, recognizing the coexistence of multiple methods and epistemologies, but also existences of the object under investigation. In doing so, I would like to argue for a displacement from the question of the mode of citizen science to the possibly larger question of the mode of existence of the objects of citizen science. I hope that framing the contours of these modes of existence will allow the emergence of coherent pragmatic and epistemic assemblages, explicit and labeled modalities of existence, that will help, in return, to clarify the need for extending the articulations of modes of citizen science.

As the discussions in the aftermath of the Tōhoku earthquake and tsunami and Fukushima Daiichi Nuclear disaster made explicit, studying disasters with the tools of academic knowledge is a humbling experience. Not a single discipline or method can describe or explain the entire chain of reactions leading to a catastrophe of the scope of the 2011 events (Fortun et al., 2016; Mazel-Cabasse, 2017, 2018). Rather, the catastrophic event can be approached as an association of distinctive modalities, or modes of knowing, that crosses the traditional divide of academic disciplines and methods: "[D]isasters come into existence in both the material and the social world and perhaps, in some hybrid space between them" (Oliver-Smith & Hoffman, 1999: 24). What seems coherent and valid from the perspective of the event is sometimes hard to articulate and prone to debate from the perspective of well-defined academic disciplines.

Science and technology scholars have noted that nineteenth-century scientists have "take[n] out the human element from the research, to make the research processes and products objective" (Strong, 2008), thus making the multiple agency, the actants, the mode of existence, that "interfere" with the scientific process invisible and, in the same movement, taken away the complexity of the subjectivity of the scientist as a knowing subject (Houdart, 2008; Mialet, 2012a). In my research, I have observed that the everyday company of earthquakes, even those yet to be, is an experience strong enough to hybridize knowledge and change the nature of expertise. A century after the 1906 earthquake, this interpretive move is still a work in progress as contemporary seismologists continue to revisit past events, and engage with scientific communities and the public. In this chapter, I have tried to show how the distinct existences of the earthquake require distinct assemblages of collaboration between scientists and non-scientists, using different types of data, and infrastructures to emerge.

In recent years, as scientific infrastructure has evolved, data used by earthquake scientists, observers, and concerned citizens have come from disparate systems of measurements: each system allows us to look at the movement of tectonic plates from particular angles and perspectives. Quantification by means of instrumentation has often taken precedence over eyewitness perceptions but, despite this considerable progress, earthquakes remain hard to grasp, and are calling for other modes of existence, other assemblages. In a world that has become increasingly datafied, where the conditions of knowledge-making are transformed to the point where researchers start evoking a change of paradigm for their discipline (Hey, Tansley, & Kristin, 2009), the questions of the modes of sciences, redistribution of research, and methods become crucial articulation to observe, analyze, and, in some cases, critique. In the context of disasters, the focus on the collective good (for the preparedness and preservation of the multiple cities of the Bay Area) has (in some places) brought about a shifting of hierarchies of knowledge. Scientists now allow themselves to become amateurs, paying attention to several modes of existence—several ontologies—of the earthquake and earthquake risk as they emerge in a situated manner. More research needs to be conducted to understand how the interaction between data structures, infrastructures, institutions, on the one hand, and workflow and repertoires, on the other, enables or constrains the emergence of new modes of existence. In this context, datafication—often thought of as a quantitative approach—must be integrated to clarify the messiness of heterogeneous data, or whatever else may be given by experience.

Bibliography

ABAG. (2013a). *Resilience initiative—Building a disaster resilient Bay Area.* Retrieved from http://quake.abag.ca.gov/projects/resilience_initiative/.

ABAG. (2013b). *Soft story buildings.* Retrieved from http://quake.abag.ca.gov/housing/softstory/.

Abe, Y. (2013). *Why safecast matters: A case study in collective risk assessment.* A STS forum on Fukushima. Berkeley. Retrieved from http://fukushimaforum.wordpress.com.

Academic Contribution: John A. Blume Earthquake Engineering Center. (n.d.). Retrieved January 1, 2017, from https://quake06.stanford.edu/centennial/tour/stop9.html.

Adger, N. (2000). Social and ecological resilience: Are they related? *Progress in Human Geography, 24*(3), 347–364. https://doi.org/10.1191/030913200701540465.

Adler, P. (1992). *Fire in the hills.* Auto-published.

Ahn, J., Carson, C., Jensen, M., Juraku, K., Mizokami, S., & Kumagai, Y. (2015). *Reflections on the Fukushima Daiichi nuclear accident.* Heidelberg and New York: Springer. https://doi.org/10.1007/978-3-319-12090-4.

Ahn, J., Guarnieri, F., & Furuta, K. (2017). *Resilience: A new paradigm of nuclear safety: From accident mitigation to resilient society facing extreme situations.* Heidelberg and New York: Springer. https://doi.org/10.1007/978-3-319-58768-4.

Akera, A. (2007). Constructing a representation for an ecology of knowledge: Methodological advances in the integration of knowledge and its various contexts. *Social Studies of Science, 37*(3), 413–441. https://doi.org/10.1177/0306312706070742.

C. Mazel-Cabasse, *Waiting for the Big One*,
https://doi.org/10.1007/978-3-030-15289-5

Akera, A., & Mohsin, A. (2016, June). Finding a place for engineering studies in disaster STS? Creating the STS forum on the 2011 East Japan disaster, 8629. https://doi.org/10.1080/19378629.2014.979830.

Alinsky, S. (1989). *Rules for radicals*. New York: Vintage; a edition.

Allen, R. (2011, April). Second before the big one. *Scientific American*, pp. 74–79. Retrieved from http://131.215.66.40/wp-content/uploads/2012/04/AllenEEWScientificAmericanApr2011_reduce.pdf.

Alquist-Priolo Earthquake Fault Zoning (AP) Act. (2012). Retrieved from http://www.conservation.ca.gov/cgs/rghm/ap/Pages/index.aspx.

Amador, F. (2004). The causes of 1755 Lisbon earthquake on Kant. *Actas VIII Congreso de La Sociedad Espanola de Historia de Las Ciencias Y de Los Tecnicos*, 485–495. Retrieved from http://dialnet.unirioja.es/servlet/dcfichero_articulo?codigo=1090088.

American Red Cross, B. A. C., Association of Bay Area Governments, California Earthquake Authority, California Geological Survey, Earthquake Engineering Research Institute, Governor's Office of Emergency Services, … U.S. Geological Survey. (n.d.). *Putting down roots in earthquake country—Your handbook for the San Francisco Bay region*. Retrieved July 25, 2012, from http://pubs.usgs.gov/gip/2005/15/.

Amin, A., & Thrift, N. (2007). Cultural-economy and cities. *Progress in Human Geography, 31*(2), 143–161. https://doi.org/10.1177/0309132507075361.

Amir, S. (2018). *The sociotechnical constitution of resilience: A new perspective on governing risk and disaster*. London and New York: Palgrave Macmillan.

Amir, S., & Kant, V. (2017). Sociotechnical resilience: A preliminary concept. *Risk Analysis, 0*(0). https://doi.org/10.1111/risa.12816.

Anderson, B., & Harrison, P. (2010). *Taking-place: Non-representational theory and geography*. Surrey: Ashgate.

AP. (2013). Gov. Brown signs bill to develop California earthquake early warning system. *The Washington Post*. Retrieved from http://articles.washingtonpost.com/2013-09-24/national/42352886_1_warning-system-earthquake-u-s-geological-survey.

Aronova, E. (2017). Citizen seismology, Stalinist science, and Vladimir Mannar's cold wars. *Science, Technology, & Human Values, 42*(2), 226–256. https://doi.org/10.1177/0162243916687643.

Arroyo, N. (2013). *No one wants to go first: S.F.'s retrofit timeline*. Retrieved from http://sfpublicpress.org/news/2012-12/no-one-wants-to-go-first-sfs-retrofit-timeline.

Arroyo, N., & Grady, B. (2013). *Earthquake retrofit delays leave thousands at risk*. Retrieved from http://sfpublicpress.org/news/2012-12/earthquake-retrofit-delays-leave-thousands-at-risk.

Atkinson, G., & Wald, D. (2007). "Did you feel it?" Intensity data: A surprisingly good measure of earthquake ground motion. *Seismological Research*

Letters, 78(3), 362–368. Retrieved from http://earthquake.usgs.gov/research/pager/prodandref/AtkinsonWaldDYFI.pdf.

Audi, T. (2014, May 19). Several earthquakes are bigger threat to San Francisco Bay Area. Multiple major temblors, not a single huge one, pose a bigger risk, study says. *Wall Street Journal*. Retrieved from http://online.wsj.com/news/articles/SB10001424052702304422704579571972490977390.

Barbalet, J. (2007). Classical pragmatism, classical sociology: William James, religion and emotion. In P. Baert & B. Turner (Eds.), *Pragmatism and European social theory* (pp. 20–21). Oxford: The Bardwell Press.

Beck, U. (1992). *Risk society, toward a new modernity* (M. Ritter, Ed.), *Nation* (Vol. 2). Sage. https://doi.org/10.2307/2579937.

Beck, U. (2008). *World at risk. New York* (Vol. The Report). Polity Publisher. Retrieved from http://scholar.google.com/scholar?hl=en&btnG=Search&q=intitle:World+at+risk:+The+report+of+the+Commission+on+the+Prevention+of+WMD+Proliferation+and+Terrorism#0.

Beck, U. (2009). Critical theory of world risk society: A cosmopolitan vision. *Constellations, 16*(1), 3–22.

Bennett, J. (2004). The force of things: Steps toward an ecology of matter. *Political Theory, 32*(3), 347–372. https://doi.org/10.1177/0090591703260853.

Benveniste, E. (1973). Problems in general linguistics. In *Problems in general linguistics* (p. 317). Miami: University of Miami Press.

Berglund, B. (2007). *Making San Fracisco America: Cultural frontiers in the urban west, 1846–1906*. Lawrence: University Press of Kansas.

Berkeley Earthquake Hits During "Shakeout" Drill. (2011, October 20). *Los Angeles Time*. Retrieved from http://latimesblogs.latimes.com/lanow/2011/10/berkeley-earthquake-hits-during-shakeout-drill.html.

Berkeley Earthquake: Much Chatter About 3.9 Tremblor, but Little Damage. (2011, October 20). *L.A. Now*. Retrieved from http://latimesblogs.latimes.com/lanow/2011/10/berkeley-earthquake-much-chatter-little-damage-about-39-temblor.html.

Bessy, C., & Chateauraynaud, F. (1995). *Experts et Faussaires*. Paris: Métailié.

Bingham, N., & Thrift, N. (2000). Thinking space. In M. Crang & N. Thrift (Eds.), *Thinking space* (pp. 281–301). London: Routledge.

Bird, E. A. R. (1987). The social construction of nature: Theoretical approaches to the history of environmental problems. *Environmental Review, 11*(4), 255–264.

Bloom, J. (2014). *An early warning device of earthquakes (and other maladies) for everyone*. Retrieved from http://5nf5.blogspot.com/2014/09/early-warning-device-of-earthquakes-and-other-maladies-for-everyone.html.

Blume, J. A. (n.d.). *John A. Blume—Trailblazing earthquake engineer*. Retrieved from https://engineering.stanford.edu/about/heroes/john-blume.

Boix Mansilla, V., Lamont, M., & Sato, K. (2016). Shared cognitive–emotional–interactional platforms: Markers and conditions for successful interdisciplinary collaborations. *Science Technology and Human Values, 41*(4), 571–612. https://doi.org/10.1177/0162243915614103.

Boltanski, L., & Thevenot, L. (2006). *On justification: Economy of worth*. Princeton: Princeton University Press.

Bonanno, G. (2004). Loss, trauma, and human resilience: Have we underestimated the human capacity to thrive after extremely aversive events? *American Psychologist, 59*(1), 20–28.

Bondi, L. (2005). Making connections and thinking through emotions: Between geography and psychotherapy. *Transactions of the Institute of British Geographers, 30*(4), 433–448. https://doi.org/10.1111/j.1475-5661.2005.00183.x.

BONDS. 2010. *Proposition A: Earthquake retrofit bond*. San Francisco County, CA.

Bonowitz, D. (2009). *The dilemma of existing buildings: Private property, public risk*. Retrieved from http://www.spur.org/publications/library/report/dilemma-existing-buildings.

Borgman, C. (2013). Scholarship in the networked world : Big data, little data, no data. In *Oliver Smithies Lecture*. Balliol College, Oxford.

Borraz, O. (2008). *Les Politiques du Risque* (Sciences P). Paris: Les Presses de Sciences Po.

Bossu, R., Gilles, S., Mazet-Roux, G., Roussel, F., Frobert, L., & Kamb, L. (2011). Flash sourcing, or rapid detection and characterization of earthquake effects through website traffic analysis. *Annals of Geophysics, 54*(6), 716–727. https://doi.org/10.4401/ag-5265.

Boudia, S., & Jas, N. (2007). Risk and risk society in historical perspective. *History and Technology, 4*, 317–331.

Bourg, D., Joly, P.-B., & Kaufmann, A. (2013). *Du risque à la Menace* (D. Bourg, P.-B. Joly, & A. Kaufmann, Eds.) (Ecologie e). Paris: Presses Universitaires de France.

Bourque, L., & Russell, L. (1994). *Experiences during and responses to the Loma Prieta earthquake*. Sacramento: California Seismic Safety Commission.

Bradford, D. (2006). *Earthquake: The other insurance crisis*.

Brady, J., & Eisner, R. (1986). *Marketing earthquake preparedness: Community campaigns that get results*. Diane Publishing.

Brechin, G. (2006). *Imperial San Francisco, urban power, earthly ruin*. Berkeley: University of California Press.

Breviglieri, M. (2006). Penser l'habiter, estimer l'habitabilité. *Tracés, 23*, 9–14.

Brown, S. (2002). Michel Serres: Science, translation and the logic of the parasite. *Theory, Culture & Society, 19*(3), 1–27.

Burby, R., & May, P. (1998). *Making building codes an effective tool for earthquake hazard mitigation* (College of Urban and Public Affairs

(CUPA)—Working Paper, 1991–2000, Paper 3). Retrieved from http://scholarworks.uno.edu/cgi/viewcontent.cgi?article=1002&context=cupa_wp.

Burress, C. (1998). *Quaking at Berkeley—New study of Northern Hayward fault forces UC to revise damage estimates Read more.* Retrieved from http://www.sfgate.com/news/article/QUAKING-AT-BERKELEY-New-study-of-northern-3008635.php#ixzz2NyfDL5Cz.

Burton, I., Kates, R., & Whites, G. (1993). *The environment as hazard* (2nd ed.). New York: The Guilford Press.

Cabane, L. (2012). *Gouverner les catastrophes. Politiques, savoirs et organisation de la gestion des catastrophes en Afrique du Sud.* CSO.

California Seismic Safety Commission. (2001). *Findings and recommendations on hospital seismic safety.* Sacramento: State of California Seismic Safety Commission.

Callon, M. (1986). Some elements of a sociology of translation: Domestication of the scallops and the fishermen of Saint-Brieuc Bay. In J. Law (Ed.), *Power, action and belief: A new sociology of knowledge?* (pp. 196–223). London: Routledge.

Callon, M. (1997). *Actor-network theory—The market test.* Actor Network and After' Workshop.

Callon, M., Lascoumes, P., & Barthe, Y. (2009). *Acting in an uncertain world: An essay on technological democracy.* Cambridge: MIT Press.

Callon, M., & Law, J. (2005). On qualculation, agency, and otherness. *Environment and Planning D: Society and Space, 23*(5), 717–733. https://doi.org/10.1068/d343t.

Callon, M., & Ferrary, M. (2006). Les réseaux sociaux à l'aune de la théorie de l'acteur-réseau. *Sociologies Pratiques, 2*(13), 37–44. https://doi.org/10.3917/sopr.013.0037.

Callon, M., Méadel, C., & Rabeharisoa, V. (2000). L'économie des qualités. *Politix, 13*(52), 211–239. https://doi.org/10.3406/polix.2000.1126.

Cantelli, F., Pattaroni, L., Roca, M., & Stavo-Debauge, J. (2009). Sensibilités pragmatiques. Enquêter sur l'action publique—Introduction. In *Sensibilités pragmatiques. Enquêter sur l'action publique* (p. 26). Bruxelles: Peter Lang.

Chakos, A. (2006). *Message to locals in California disasters: YOYO-You're on your own.*

Chakos, A. (2010). Getting a jump on the next big disaster. *Huffington Post.* Retrieved from http://www.huffingtonpost.com/arrietta-chakos/getting-a-jump-on-the-nex_b_588990.html?.

Chakos, A. (2011). *Toward a resilient future: A review of Palo alto's emergency readiness.*

Chakos, A., Schulz, P., & Tobin, L. T. (2002). Making it work in Berkeley: Investing in community sustainability. *Natural Hazard Review, 3*, 55–67.

Chang, K.-K., Salmon, C., Lee, B., Choi, J., & Zeldes, G. (2010). The influence of contextual factors on the selection of news frames—A Cross-National approach to the news coverage of severe acute respiratory syndrome (SRAS).

In G. Golan, T. Jonhson, & W. Wanta (Eds.), *International media communication in a global age* (p. 488). London: Routledge.
Charvolin, F. (1997). La Naissance du Domaine de l'Environnement. *Strates [Online], 9*. Retrieved from http://strates.revues.org/636.
Chateauraynaud, F. (2008). Les mobiles de l'expertise Entretien avec Francis Chateauraynaud. *Experts, 78*, 4.
Chavrolin, F. (2013). Pense-bêtes, astuces et recettes de jardiniers-observateurs de papillons. Retour sur une science citoyenne. *Revue d'Anthropologie Des Connaissances, 7*(2), 485–500.
Clark, L. (2006). *Worst cases, terror and catastrophe in the popular imagination*. Chicago: University of Chicago Press.
Clavandier, G. (2004). *La mort collective* (CNRS Socio). Paris: CNRS Editions.
Coen, D. (2013). *The earthquake observers: Disaster science from Lisbon to Richter*. Chicago: The University of Chicago Press.
Collectif. (2008a). Catastrophe(s)? *Le Portique, 2ème semes*(22).
Collectif. (2008b). *Le temps des Catastrophes* (M.-O. Padis, Ed.), *Esprit* (Esprits, Vol. Mars–Avril).
Comfort, L. K. (2006). City at risk: Hurricane Katrina and the drowning of New Orleans. *Urban Affairs Review, XX*(X), 1–16. https://doi.org/10.1177/1078087405284881.
Comfort, L. K., Boin, A., & Demchak, Chris, C. (2010). *Designing resilience: Preparing for extreme events* (1st ed.). Pittsburgh: University of Pittsburgh Press.
Creton-Cazanave, L. (2010). *Penser l'Alerte par les Distances. Entre Planification et Emancipation, l'Exemple du Processus d'Alerte aux Crues Rapides sur le Bassin Versant du Vidourle*. Université de Grenoble.
Cronon, W. (1992). *Nature's metropolis, Chicago and the great West*. New York: Norton.
Curtis, G. H. (2008). *Emeritus professor of geology weighs in on memorial stadium*. The Berkeley Daily Planet.
Cutter, S. (2005). Are we asking the right questions? In R. Perry & E. L. Quarantelli (Eds.), *What is a disaster?: New answer to old questions* (pp. 39–49). Philadelphia: Xlibris.
Daston, L. (1991). Marvelous facts and miraculous evidence in early modern Europe. *Critical Inquiry, 18*(1), 93–124. Retrieved from http://www.jstor.org/stable/1343717.
D'Ercole, R., Gluski, P., Hardy, S., & Sierra, A. (2009). Vulnérabilités urbaines dans les pays du sud. *Cybergeo: Journal of Geography, 1*, 95–98.
Datson, L., & Galison, P. (2010). *Objectivity*. Cambridge: Zone Books.
Davidson, J., Bondi, L., & Smith, M. (2005). *Emotional geographies*. Aldershot: Ashgate.
Davis, M. (1998). *Ecology of fear: Los Angeles and the imagination of disasters*. New York: Vintage Books.

de Certeau, M. (1988). *The practice of everyday life*. Berkeley: University of California Press.

Deep Under Berkeley. (2011). Retrieved from http://seismo.berkeley.edu/blog/seismoblog.php/2011/10/20/deep-under-berkeley.

Deleuze, G. (1987). *Leibniz. Logique de l'évènement*. Retrieved from www.webdeleuze.com.

Demeritt, D. (2002). What is the "social construction of nature"? A typology and sympathetic critique. *Progress in Human Geography, 26*(6), 767–790. https://doi.org/10.1191/0309132502ph402oa.

Demuth, J. L. (2002). Countering terrorism: Lessons learned from natural and technological disasters. In *National Academy of Sciences* (p. 36). Washington, DC.

Depraz, N. (2013). Attention et Affection: la micro-genèse husserlienne de l'attention à la lumière des perspectives empiriques de Stumpf, James, de Külpe et de Titchener. In *Phenomenology 2005, Vol. 3: Selected essays from Euro-Mediterranean area, Part 1* (English, Spanish and French Edition) (pp. 103–135). Bucharest: Zeta Books.

Dosse, F. (1995). *L'Empire du Sens. L'Humanisation des Sciences Humaines*. Paris: La Decouverte.

Douglas, M., & Wildavsky, A. (1983). *Risk and culture: An essay and the selection of technological environmental dangers*. Berkeley: University of California Press.

Dupuy, J.-P. (2002). *Pour un catastrophisme éclairé. Quand l'impossible est certain* (Point Essa). Paris: Seuil.

Dynes, R. R. (1971). *The dialogue between Voltaire and Rousseau on the Lisbon earthquake: The emergence of social science view* (No. DE 19716).

Dynes, R. R. (1997). *The Lisbon earthquake in 1755: Contested meaning in the first modern disaster* (No. #255).

Earthquake Rocks Haiti. (2010). Retrieved from http://thehaitianblogger.blogspot.com/2010/01/earthquake-70-to-73-rocks-haiti.html.

Eitzel, M. V, Cappadonna, J. L., Santos-Lang, C., Duerr, R. E., Virapongse, A., West, S. E., … Jiang, Q. (2017). Citizen science terminology matters: Exploring key terms. *Citizen Science: Theory and Practice, 2*(1), 1. https://doi.org/10.5334/cstp.96.

Fante, J. (1939). *Ask the dust* (Modern Cla). New York: HarperPerenial.

Farmer, P. (2011). *Haiti after the earthquake*. New York: PublicAffairs.

Favier, R., & Granet-Abisset, A.-M. (2009). Society and natural risk in France, 1500–2000: Changing historical perspectives. In C. Mauch & C. Pfister (Eds.), *Natural disasters cultural response* (pp. 103–136). Lanham: Lexington Book.

Filson, J., McCarthy, J., Ellsworth, W., & Zorback, M.-L. (2003). *The USGS earthquake hazard program in NEHRP—Investing in a safer future*. Retrieved from http://pubs.usgs.gov/fs/2003/fs017-03/.

Flichy, P. (2008). *The internet imaginaire*. Cambridge: MIT Press.

Folke, C. (2006). Resilience: The emergence of a perspective for social-ecological systems analyses. *Global Environmental Change, 16*(3), 253–267. https://doi.org/10.1016/j.gloenvcha.2006.04.002.

Fortun, K., Knowles, S., Choi, V., Jobin, P., Matsumoto, M., de la Torre III, P., ... Murillo, L. F. (2016). Researching disaster from an STS perspective. In U. Felt, R. Fouche, C. A. Miller, & L. Smith-Doerr (Eds.), *The handbook of science and technology studies* (4th ed., pp. 1003–1028). Cambridge, MA: MIT Press.

Foucault, M. (1970). *The order of things: An archaeology of the human sciences.* London and New York: Routledge Classics.

Frankel, H. (2012). *The continental drift controversy: Evolution into plate tectonics* (Vol. 4). Cambridge: Cambridge University Press.

Fressoz, J. (2007). Beck back in the 19th century: Towards a genealogy of risk society. *History and Technology, 23*(4), 333–350. https://doi.org/10.1080/07341510701527419.

Fressoz, J.-B. (2008). Comment sommes nous devenus modernes? Petite histoire philosophique du risque et de l'expertise à propos de l'innoculation de la vaccine, 1750–1800. In S. Topçu, S. Cuny, & K. Serrano-Velarde (Eds.), *Savoirs en débat: Perspectives Franco-Allemandes* (Vol. 39, pp. 1–24). Paris: L'Harmattan.

Fressoz, J. (2012). *L'apocalypse joyeuse : Une histoire du risque technologique.* Paris: Seuil.

Frickel, S., & Bess, V. (2007). Katrina, contamination, and the unintended organization of ignorance. *Technology in Society, 29,* 181–188.

Frickel, S., Campanella, R., & Besse, V. (2009). Mapping knowledge investments in the aftermath of Hurricane Katrina: A new approach for assessing regulatory agency responses to environmental disaster. *Environmental Science and Policy, 12*(2), 119–133.

Gaar, G., & Miller, R. W. (2006). *San Francisco: A natural history* (Image of A). San Francisco: Arcadia Publishing.

Galin, D. (1994). The structure of awareness: Contemporary applications of William James' forgotten concept of fringe. *Journal of Mind and Behavior, 15*(4), 375–400.

Genthes, A. (1936). *As I remember.* Retrieved from http://www.zpub.com/sf/history/ag/ag-asi-10.html.

Gerson, E. M. (2012). Integration of specialties: An institutional and organizational view. *Studies in History and Philosophy of Science Part C: Studies in History and Philosophy of Biological and Biomedical Sciences, 44*(4), 515–524. https://doi.org/10.1016/j.shpsc.2012.10.002.

Geschwind, C.-H. (2001). *California earthquake, science, risk and the politic of hazard mitigation.* Baltimore and London: The Johns Hopkins University Press.

Gilbert, C. (2007). Crisis analysis: Between normalization and avoidance. *Journal of Risk Research, 10*(7), 925–940. https://doi.org/10.1080/13669870701504731.

Gomart, E., & Hennion, A. (1999a). A sociology of attachment: Music amateurs, drug users. *The Editorial Board of the Sociological Review*. https://doi.org/10.1111/j.1467-954X.1999.tb03490.x.

Gomart, E., & Hennion, A. (1999b). A sociology of attachment: Music amateurs, drug users. In J. Law & J. Hassard (Eds.), *Actor network theory and after* (pp. 1–28). Oxford: Wiley-Blackwell.

Gorney, C. (2009). *Remembering the surgery inside the freeway collapse*. Retrieved from http://oaklandnorth.net/2009/10/17/remembering-the-surgery-inside-the-freeway-collapse/.

Gralepois, M. (2008). *Les risques collectifs dans les agglomérations françaises. Contours et limites d'une approache territoriale de prévention et de gestion des risques à travers le parcours des agents administratifs locaux*. Universite de Paris Est.

Greenberg, J. (2010). *From Betamax to blockbuster: Video stores and the invention of movies on video*. Cambridge: MIT Press.

Grossi, P., & Muir-Wood, R. (2006). *The 1906 San Francisco earthquake and fire: Perspectives on a modern super cat*. San Francisco.

Grossi, P., & Zoback, M. Lou. (2009). *Catastrophe modeling and California earthquake risk: A 20-year perspective*.

Grossi, P., & Zoback, M. L. (2010). *1868 Hayward earthquake: 140 year retrospective*. Menlo Park: Risk Management Solutions.

Groth, P., & Bressi, T. (1997). *Understanding ordinary landscapes*. New Haven: Yale University Press.

Guarnieri, F. (2017). The Fukushima Daiichi nuclear accident: Entering into resilience faced with an extreme situation. In J. Ahn, F. Guarnieri, & K. Furuta (Eds.), *Resilience: A new paradigm of nuclear safety: From accident mitigation to resilient society facing extreme situations* (pp. 1–17). Heidelberg and New York: Springer. https://doi.org/10.1007/978-3-319-58768-4.

Guarnieri, F., & Travadel, S. (2018). *Un récit de Fukushima. Le directeur parle*. Paris: Presses Universitaires de France.

Guggenheim, D. (2006). *An inconvenient truth*. Los Angeles, USA: Paramount Classic.

Hall, P. A., & Lamont, M. (2013). Introduction. In P. A. Hall, & M. Lamont (Eds.), *Social resilience in the neoliberal era* (pp. 1–34). New York: Cambridge University Press.

Haas, P. (1992). Epistemic communities and international policy coordination. *International Organization, 46*(1), 1–35.

Hache, E. (2011). *Ce à quoi nous tenons. Propositions pour une écologie pragmatique* (Les empech). Paris: La Decouverte.

Hackett, E., Amsterdamska, O., Lynch, M., & Wajcman, J. (Eds.). (2008). *The handbook of science and technology studies*. Cambridge: MIT Press.

Haiti: Earthquake! (2010). Retrieved from http://globalvoicesonline.org/2010/01/13/haiti-earthquake/.

Hansen, G. (1989). *Denial of disaster: The untold story and photographs of the San Francisco earthquake of 1906.* San Francisco: Cameron & Company.

Hansson, S. O. (2012). Risk. In S. O. Hansson (Ed.), *The Stanford encyclopedia of philosophy.* Stanford: Stanford University. https://plato.stanford.edu/entries/risk/.

Haraway, D. (1988). Situated knowledges: The science question in feminism and the privilege of partial perspective. *Feminist Studies, 14*(3), 575–599.

Haring, K. (2006). *Ham radio's technical culture.* Cambridge: The MIT Press.

Harman, G., Bryant, L., & Srnicek, N. (Eds.). (2009). *The speculative turn continental materialism and realism* (Anamnesis, pp. 304–334). Melbourne: re.press. Retrieved from http://books.google.com/books?id=coBDqJQeAQYC&pg=PA304&lpg=PA304&dq=Latour+Souriau&source=bl&ots=wHqRXjOttu&sig=GjPBBQjrUsrgrRgvWeykV9m6Q7Q&hl=en&sa=X&ei=XDFnUe_ZNoXmiwK7toDADQ&ved=0CGIQ6AEwAw#v=onepage&q=LatourSouriau&f=false.

Harper, Dan. (2010). *Creation myth of Northern Ohlone (Coastonan) Indian.*

Harvey, D. (1997). *Justice, nature and the geography of difference.* Oxford: Blackwell.

Hasegawa, R. (2013). *Disaster evacuation from Japan's 2011 tsunami disaster and the Fukushima nuclear accident—Studies no. 05/13.* Paris. Retrieved from https://www.iddri.org/sites/default/files/import/publications/study0513_rh_devast-report.pdf.

Heller, A. (2006). *Re-creating the 1906 San Francisco earthquake.* Retrieved from https://www.llnl.gov/str/Sep06/Rodgers.html.

Hennion, A. (2004a). Pragmatics of taste. In M. Jacobs & N. Hanrahan (Eds.), *The Blackwell companion to the sociology of culture* (pp. 131–144). Oxford: Blackwell.

Hennion, A. (2004b). Une sociologie des attachements. D'une sociologie de la culture à une pragmatique de l'amateur. *Sociétés. Pratiques Musicales, 3*(85), 9–24.

Hennion, A. (2007). Those things that hold us together: Taste and sociology. *Cultural Sociology, 1*(1), 97–114. https://doi.org/10.1177/1749975507073923.

Henry, D. (2005). Anthropology and disasters. In D. McEntire & W. Blanchard (Eds.), *Disciplines, disasters and emergency management: The convergence and divergence of concepts, issues and trends from the research literature* (p. 1). Federal Emergency Management Agency. https://doi.org/10.2307/3032499.

Hey, T., Tansley, S., & Kristin, T. (2009). *The fourth paradigm: Data-intensive scientific discovery: Microsoft research.* Redmond, WA. Retrieved from https://www.immagic.com/eLibrary/ARCHIVES/EBOOKS/M091000H.pdf.

Hilton, S. (2012, February). Christchurch finding its way back from devastation. *San Francisco Chronicle*. Retrieved from http://www.sfgate.com/travel/article/Christchurch-finding-its-way-back-from-devastation-3252789.php.

Hinman, E., & Hutchinson, D. A. (2005). *San Francisco's earthquake risk*.

Hoffman, S. (1994). Up from the embers: A disaster survivor's story. *NCP Clinical Quarterly, 4*(2), 15–16.

Hoffman, S. (1998). Eve and Adam among the embers: Gender patterns after the Oakland Berkeley firestorm. In E. Enarson & B. Hern Morrow (Eds.), *The gendered terrain of disaster: Through women's eyes* (pp. 55–61). Westport, CT: Greenwood Publishing.

Hoffman, S. M., & Oliver-Smith, A. (2002). *Catastrophe and culture: The anthropology of disasters* (S. M. Hoffman & A. Oliver-Smith, Eds.). School of American Research Press and James Currey.

Holden. (2011). *How much money has been given and spent for Haiti earthquake relief? Putting the numbers in perspective*. Retrieved from How much money has been given and spent for Haiti earthquake relief? Putting the numbers in perspective.

Houdart, S. (2008). *La cour des miracles: Ethnologie d'un laboratoire japonais* (Sociologie). Paris: CNRS Editions.

Houdart, S., & Thiery, O. (2011). *Humains non humains. Comment repeupler les sciences sociales?* Paris: La Découverte.

Hutchings Mieler, D., & Brechwald, D. (2013). *Regional resilience initiative*.

Imaoka, L. (2013). Mediated maps: Geographic interoperability issues following global disasters. In *Society for Social studies of Sciences*. San Diego.

Ingold, T. (2005). *Learning through doing and understanding in practices*. Retrieved from http://www-new1.heacademy.ac.uk/assets/Documents/subjects/csap/Learning_though_doing_and_understanding_in_practice.pdf.

Inventory of the Joint Committee on Seismic Safety Records. (n.d.). Retrieved January 12, 2017, from http://www.oac.cdlib.org/findaid/ark:/13030/tf0290005q/.

Issues Answer. (2009). *San Francisco Bay Area preparedness—Executive summary*.

Jaded West Coast chuckles Over East Coast Quake. (2011). Retrieved from http://abclocal.go.com/kfsn/story?section=news/national_world&id=8323344.

James, W. (1890). *Classics in the history of psychology. Henry Holt and Company*. Retrieved from https://psychclassics.yorku.ca/James/Principles/prin11.htm.

James, W. (1892). *Psychology: The brief course*. Toronto: Dover Publication.

James, W. (1906). On some mental effects of the earthquake. *The Youth's Companion*. Reprinted in James, H. Jr., 1911, *Memories and studies* (H. James, Jr.). Longmans, Green, & Co. (Vol. June). New York.

Jasanoff, S. (1986). *Risk management and political culture* (Social Res). New York: Russell Sage Foundation.

Jasanoff, S. (2010). Beyond calculation: A democratic response to risk. In *Disaster and the politic of intervention* (pp. 14–41). New York: Columbia University Press.

Jasanoff, S. (2014). Genealogies of STS. *Social Studies of Science, 42*(3), 435–441. https://doi.org/10.1177/0306312712440174.

Jasanoff, S., & Kim, S. (2015). *Dreamscapes of modernity: Sociotechnical imaginaries and the fabrication of power*. Chicago: The University of Chicago Press.

Jervollino, D. (2006). *The future of phenomenology: Towards a philosophy of translation inspired by a phenomenological hermeneutics*. Retrieved from http://www.newschool.edu/nssr/husserl/Future/PartOne/Jervolino.html.

Jirotka, M., Lee, C. P., & Olson, G. M. (2013). Supporting scientific collaboration: Methods, tools and concepts. *Computer Supported Cooperative Work: CSCW: An International Journal, 22*(4–6), 667–715. https://doi.org/10.1007/s10606-012-9184-0.

Johnson, L. (2011). *Insuring climate change: Science, fear, and value in reinsurance markets*. Berkeley: University of California.

Jonas, H. (1995). *Le principe de responsabilité* (Champs Ess). Paris: Flammarion.

Jones, C. (2007, February 20). Tree-sitters say site near stadium may be Native American burial ground. *San Francisco Chronicle*. Retrieved from http://www.sfgate.com/bayarea/article/BERKELEY-Tree-sitters-say-site-near-stadium-may-2647672.php.

Jones, C. (2008, September 25). Retrofit plan to ride out quake at Cal stadium. *SFGates*. Retrieved from https://www.sfgate.com/bayarea/article/Retrofit-plan-to-ride-out-quake-at-Cal-stadium-3193674.php.

Juraku, K. (2015). Why is it so difficult to learn from accidents? In *International Workshop on Nuclear Safety: From Accident Mitigation to Resilient Society Facing Extreme Situations*. Berkeley.

Kalmbach, K. (2015). From Chernobyl to Fukushima: The impact of the accidents on the French nuclear discourse. In T. Bohn & et.al (Eds.), *The Impact of disaster: Social and cultural approaches to Fukushima and Chernobyl* (pp. 67–96). Berlin: EB Verlag. Retrieved from https://research.tue.nl/en/publications/from-chernobyl-to-fukushima-the-impact-of-the-accidents-on-the-fr.

Karlamangla, S. (2011). *Eight earthquakes hit Berkeley over past several days*.

Keeley, J. (2005). Fire history of the San Francisco East Bay region and implications for landscape patterns. *International Journal of Wildland Fire, 14*, 285–296.

Kimura, A., & Kinchy, A. (2016). Citizen science : Probing the virtues and contexts of participatory research. *Engaging Science, Technology and Society, 2*, 331–361. https://doi.org/10.17351/ests2016.099.

King, J. (2011). *Cityscapes: San Francisco and its buildings*. Berkeley: Heyday.

Kingston, M. H. (2003). *The fifth book of peace*. New York: First Vintage International Edition.

Kircher, C. A., Seligson, H. A., Bouabid, J., & Morrow, G. C. (2006). When the big one strikes again—Estinated losses due to a repeat of the 1906 San Francisco earthquake. *Earthquake Spectra, 22*(2), 297–339.

Knobel, L. (2014). Cal prof Joshua Bloom makes earthquake alert kit for $110. Retrieved from http://www.berkeleyside.com/2014/09/09/uc-berkeley-prof-joshua-bloom-makes-earthquake-alert-kit-for-110/.

Knorr-Cetina, K. (1999). *Epistemic cultures: How the sciences make knowledge.* Cambridge, MA: Harvard University Press. Retrieved from https://books.google.com/books?id=WFEeib0Q9L0C&pgis=1.

Knowles, S. (2011). *The disaster experts: Mastering risk in modern America.* Philadelphia: The University of Pennsylvania Press.

Kullenberg, C., & Kasperowski, D. (2016). What is citizen science?—A scientometric meta-analysis. *PloS ONE, 11*(1), e0147152. https://doi.org/10.1371/journal.pone.0147152.

Lagadec, P. (1987). Le risque technologique majeur. In *Universalia.* Paris: Encyclopedia Universalis.

Lamy, A. (2007). Récits médiatiques, mémoires électives. *EspacesTemps.Net.* Retrieved from http://espacestemps.net/document3673.html.

Landström, C., Whatmore, S. J., & Lane, S. N. (2011). Virtual engineering: Computer simulation modelling for flood risk management in England. *Science Studies, 24*(2), 3–22.

Lane, S., Odoni, N., Landstrom, C., Whatmore, S., Ward, N., & Bardley, S. (2010). Doing flood risk science differently: An experiment in radical scientific method. *Transaction of the Institute of British Geographers, 36,* 15–36.

Langumier, J. (2008). *Survivre à l'innondation. Pour une ethnologie de la catastrophe.* Lyon: ENS éditions.

Lash, S., Szerszynski, B., & Wynne, B. (1996). *Risk environment and modernity: Toward a new ecology* (L, Ed.) (Published). London: Sage.

Latour, B. (1991). *Nous n'avons jamais été modernes. Essai d'anthropologie symétrique* (Poche). Paris: La Découverte.

Latour, B. (1993). *The pasteurisation of France.* Cambridge: Harvard University Press.

Latour, B. (2004). Whose cosmos? Which cosmopolitics? A commentary on Ulrich Beck's peace proposal? *Common Knowledge, 10*(3), 450–462.

Latour, B. (2005a). What is given in experience? *Boundary 2, 32*(1), 223–237. https://doi.org/10.1215/01903659-32-1-223.

Latour, B. (2005b). What is the style if matters of concerns. In *Spinoza Lectures* (p. 27).

Latour, B. (2007). *Changer de société, refaire de la sociologie* (Poche Scie). Paris: La Découverte.

Latour, B. (2009). *Sur le Culte Moderne des Dieux Faitiches.* Paris: Les Empecheurs de Tourner en Rond, La Decouverte.

Latour, B. (2011). Reflections on Etienne Souriau's Les differents modes d'existence. In G. Harman, L. Bryant, & N. Srnicek (Eds.), *The speculative turn continental materialism and realism* (Anamnesis, pp. 304–334). Melbourne: re.press. Retrieved from http://books.google.com/books?id=coB-DqJQeAQYC&pg=PA304&lpg=PA304&dq=Latour+Souriau&source=bl&ots=wHqRXjOttu&sig=GjPBBQjrUsrgrRgvWeykV9m6Q7Q&hl=en&sa=X&ei=XDFnUe_ZNoXmiwK7toDADQ&ved=0CGIQ6AEwAw#v=onepage&q=LatourSouriau&f=false.

Latour, B. (2012). *Enquête sur les modes d'existence. Une anthropologie des Modernes.* Paris: La Découverte.

Latour, B. (2013). habit [hab], too, is a mode of existence in *An inquiry into modes of existence.* Cambridge, MA: Harvard University Press, p. 265.

Latour. (2014). *Habit.* [hab]. Retrieved May 4, 2019, from http://modesofexistence.org/crossings//#/en/hab.

Latour, B., & Girard Stark, M. (1999). Factures/fractures: From the concept of network to the concept of attachment. *RES: Anthropology and Aesthetics, 36*(Autumn), 20–31. Retrieved from http://www.bruno-latour.fr/sites/default/files/downloads/76-FAKTURA-GB.pdf.

Law, J., & Lien, M. E. (2013). Slippery: Field notes in empirical ontology. *Social Studies of Science, 43*(3), 363–378. https://doi.org/10.1177/0306312712456947.

Lazzarotti, O. (2006). *Habiter, la condition géographique* (Mapemonde). Paris: Belin.

Leonelli, S., Rappert, B., & Davies, G. (2017). Data shadows. *Science, Technology, & Human Values, 42*(2), 191–202. https://doi.org/10.1177/0162243916687039.

Lévy, J., & Lussault, M. (2000). *Logique de l'espace, esprit des lieux. Géographie à Cerisy.* Paris: Belin.

Lewis, J. (2008, October 21). *The coming quake: Is Los Angeles ready for the Big One?* High Country News. Retrieved from https://www.hcn.org/issues/40.19/the-coming-quake/?b_start:int=1.

Lidskog, R. (2008). Scientised citizens and democratised science: Re-assessing the expert-lay divide. *Journal of Risk Research, 11*(1), 69–86. https://doi.org/10.1080/13669870701521636.

Lin, R.-G. I. (2017, May 3). Earthquake early warning system nets $10.2 million in Congress' budget deal. *Los Angeles Time.* Retrieved from http://www.latimes.com/local/lanow/la-me-ln-earthquake-early-warning-20170502-story.html.

LittleJohn, D. (2011). *The big one. A story of San Francisco.* Durham: Strategic Book Group.

Livingston, A. (2012). *Excited subjects: William James and the politics of radical empiricism: Theory and event* (Vol. 15). Project MUSE. Retrieved from https://muse.jhu.edu/article/491201.

Lupton, D. (1999a). *Risk*. London: Routledge.

Lupton, D. (1999b). *Risk and sociocultural theory: New directions and perspectives*. Cambridge: Cambridge University Press.

Lussault, M. (2007). *de l'espace humain*. Paris: Editions du Seuil.

Lynch, M. (2012). Self-exemplifying revolutions? Notes on Kuhn and Latour. *Social Studies of Science, 42*(3), 449–455. https://doi.org/10.1177/0306312712439120.

Lynch, M. (2013). Ontography: Investigating the production of things, deflating ontology. *Social Studies of Science, 43*(3), 444–462. https://doi.org/10.1177/0306312713475925.

Lynch, M. (2014). Genius, Hawking, and expertise. *Social Studies of Science, 44*, 793.

Maffei, J. (2010). *The coming Bay Area earthquake, 2010 update of scenario for a magnitude 7.0 on the Hayward fault*.

Mahony, M., & Hulme, M. (2016). Epistemic geographies of climate change. *Progress in Human Geography*, 030913251668148. https://doi.org/10.1177/0309132516681485.

Main, I., Bell, A., Meredith, P., Geiger, S., & Touati, S. (2012). The dilatancy-diffusion hypothesis and earthquake predictability. *Geological Society of London, 367*, 215–230.

Margolin, M. (2003). *The Olhone Way*. Berkeley: Heyday.

Marres, N. (2012). The redistribution of methods: On intervention in digital social research, broadly conceived. *Sociological Review, 60*(Suppl. 1), 139–165. https://doi.org/10.1111/j.1467-954X.2012.02121.x.

Marres, N. (2013). Why political ontology must be experimentalized: On eco-show homes as devices of participation. *Social Studies of Science, 43*(3), 417–443. https://doi.org/10.1177/0306312712475255.

Matthews, D. (2011). *The best earthquake tweets*. Retrieved from http://www.washingtonpost.com/blogs/wonkblog/post/top-16-earthquake-tweets/2011/08/23/gIQANgmJZJ_blog.html.

Mayor Lee & Supervisors Sign City's Mandatory Seismic Retrofit Program for Soft Story Buildings. (2013). San Francisco. Retrieved from Mayor Lee & Supervisors Sign City's Mandatory Seismic Retrofit Program for Soft Story Buildings.

Mazel-Cabasse, C. (2017). Hybrid disasters—Hybrid knowledge. In J. Ahn, F. Guarnieri, & K. Furuta (Eds.), *Resilience: A new paradigm of nuclear safety: From accident mitigation to resilient society facing extreme situations* (pp. 337–351). Heidelberg and New York: Springer. https://doi.org/10.1007/978-3-319-58768-4.

Mazel-Cabasse, C. (2018). What (sociotechnical) resilience is made of: Personal trajectories and earthquake risk mitigation in the San Francisco Bay Area. In S. Amir (Ed.), *The sociotechnical constitution of resilience: a new perspective on governing risk and disaster*. London: Palgrave Macmillan.

Mialet, H. (1994). *Le Sujet de l'invention. Etude empirique de la conception d'une idee neuve: comparaison des methodes philosophique et sociologique*. Sorbonne Paris I.

Mialet, H. (2012a). *Hawking incorporated: Stephen Hawking and the anthropology of the knowing subject*. Chicago and London: University of Chicago Press.

Mialet, H. (2012b). Where would STS be without Latour? What would be missing? *Social Studies of Science*. https://doi.org/10.1177/0306312712440311.

Mileti, D. S., & DeRouen Darlington, J. (1995). Societal response to revised earthquake probabilities in the San Francisco Bay Area. *International Journal of Mass Emergencies Disasters, 13*(2), 119–145.

Miller, F., Osbahr, H., Boyd, E., Thomalla, F., Bharwani, S., Ziervogel, G., … Nelson, D. (2010). Resilience and vulnerability: Complementary or conflicting concepts? Retrieved from http://www.ecologyandsociety.org/vol15/iss3/art11/.

Min, L., & Perkins, J. B. (2008). *Long-term disaster recovery planning by local governments in the San Francisco Bay Area*.

Mitchell, T. (2002). *Rule of experts: Egypt, techno-politics, modernity*. Berkeley: University of California Press.

Mitchell, M. (2011). *Fire ruin renewal*. The Phenix Firestorm Project. Retrieved from http://www.margarettamitchell.com/firestorm/.

Mol, A. (2010). Actor-network theory: Sensitive terms and enduring tensions. *Kölner Zeitschrift Für Soziologie Und Sozialpsychologie, 50*(1), 253–269.

Mol, A. (2013). Mind your plate! The ontonorms of Dutch dieting. *Social Studies of Science, 43*(3), 379–396. https://doi.org/10.1177/0306312712456948.

Montgomery, D. (2004). *King of fish: The thousand year run of salmon*. New York: Basic Books.

Morisy, M. (2011). *A year later, lessons for the media from the Haiti earthquake response*.

Mulargia, F., & Geller, R. J. (2012). *Earthquake science and seismic risk reduction* (N. S. Series, Ed.). Dordrecht: Springer Science & Business Media.

Murakami, H. (2002). *After the quake*. New York: First Vintage International Edition.

Murdoch, J. (1998). The spaces of actor-network theory. *Geoforum, 29*(4), 357–374. https://doi.org/10.1016/S0016-7185(98)00011-6.

Myers, D., & Wee, D. F. (2005). *Disaster, mental health services: A primer for practitioners*. New York: Brunner-Routledge.

Neyrat, F. (2008). *Bio-politique des catastrophes* (Dehors). Paris: Editions MF.

November, V., Camacho-Hübner, E., & Latour, B. (2010). Entering a risky territory: Space in the age of digital navigation. *Environment and Planning D: Society and Space, 28*(4), 581–599. https://doi.org/10.1068/d10409.

November, V., & Leanza, Y. (2015). *Risk, disaster and crisis reduction*. Cham: Springer. https://doi.org/10.1007/978-3-319-08542-5.

November, V., Penelas, M., & Viot, P. (2009). When flood risk transforms a territory: The Lully effect. *Geography, 94*, 189–197.

Nye, R. (2014). *The next big step*. Retrieved from http://www.risanye.com/.

Obama, B. (2008). *Why organize?* Retrieved from http://illinoisissues.uis.edu/archives/2008/09/whyorg.html.

Oliver-Smith, A., & Hoffman, S. M. (1999). In A. Oliver-Smith & S. M. Hoffman (Eds.), *The angry earth: Disaster in anthropological perspective* (1st ed.). Oxford, UK: Routledge.

Olson, R. A. (n.d.). *Robert A. Olson*. Retrieved from http://www.olsonassoc.com/utility/about.html.

Oreskes, N., & Conway, E. (2011). *Merchants of doubt: How a handful of scientists obscured the truth on issues from tobacco smoke to global warming* (Reprint edition). New York: Bloomsbury Press.

Padilla, A. (2014). *SB-135 earthquake early warning system*. Senate.

Paquot, T., Lussalut, M., & Younes, C. (2007). *Habiter, le propre de l'humain. Villes, territoires et philosophie* (Armillaire). Paris: La Decouverte.

Pattaroni, L., & Pedrazzini, Y. (2010). Insecurity and segregation: Rejecting an urbanism of fear. In P. Jacquet, R. Pachauri, & L. Tubina (Eds.), *Cities: Steering towards sustainability* (pp. 177–187). Delhi: TERI Press.

Paxton, J. (2004, August). Earthquakes: San Francisco at risk. *Spur NewsLetter*.

Peck, R. (2013). Assistance Mortelle/Fatal Assistance. Arte France (in co-production with), Canal Overseas Productions (with the participation of), Centre National de la Cinématographie et de l'Image Animée (participation), Radio Télévision Belge Francofone (RTBF) (in co-production with), Radio Télévision Sui.

Pelling, M. (2003). Tracing the roots of urban risk and vulnerability. In *The vulnerability of cities: Natural disasters and social resilience* (p. 224). London: Routledge.

Perkins, J. B. (2005). *Taming natural disasters, multi-jurisdictional local government hazard mitigation plan for the San Francisco Bay Area*.

Perkins, J. B., Chakos, A., Olson, R. A., Tobin, L. T., & Turner, F. (2006). A retrospective on the 1906 earthquake's impact on Bay Area and California public policy. *Earthquake Spectra, 22*(S2), S237. https://doi.org/10.1193/1.2181527.

Perkins, J., & Hutchings, D. (2010). *Taming natural disasters: Multi-jurisdictional local hazard mitigation plan for the San Francisco Bay Area: 2010 update of 2005 plan*. Oakland.

Perrow, C. (2006). The disaster after 9/11: The department of homeland security and the intelligence reorganization. *Homeland Security Affairs, II* (Article 3). Retrieved from https://www.hsaj.org/articles/174%0A%0A%0A%0A.

Petersen, M. D., Frankel, A. D., Harmsen, S. C., Mueller, C. S., Haller, K. M., Wheeler, R. L., … Luco, N. (2011). *Seismic-hazard maps for the conterminous United States, 2008.*

Pigeon, P. (2005). *Géographie Critique des Risques* (Antropos). Paris: Economica.

Podestá, G. P., Natenzon, C. E., Hidalgo, C., & Ruiz Toranzo, F. (2013). Interdisciplinary production of knowledge with participation of stakeholders: A case study of a collaborative project on climate variability, human decisions and agricultural ecosystems in the Argentine Pampas. *Environmental Science and Policy, 26,* 40–48. https://doi.org/10.1016/j.envsci.2012.07.008.

Poland, C. (2009). *The resilient city: Defining what San Fracisco needs from its seismic mitigation policies.*

Post Quake, West Coast Teases East on Social Networks. (2011). Retrieved from http://sanfrancisco.cbslocal.com/2011/08/23/post-quake-west-coast-teases-east-on-social-networks/.

Pratt, M. L. (1992). *Imperial eyes: Travel writing and transculturation.* London: Routledge.

Quarantelli, E. L. (1998). *What is a disaster?: A dozen perspectives on the question.* London: Routledge.

Quarantelli, E. L. (1999). *Disaster related social behavior: Summary of 50 years of research findings.*

Quarantelli, E. L., & Perry, R. (2005). *What is a disaster.* Xlibris: Bloomington.

Quenet, G. (2005). *Les Tremblements de terre aux dix-septième et dix-huitième siècles. La naissance d'un risque.* Seyssel: Champ Vallon.

Ramachandran, V., & Walz, J. (2013). Haiti's earthquake generated a $9bn response—Where did the money go. *The Guardian.* Retrieved from http://www.guardian.co.uk/global-development/poverty-matters/2013/jan/14/haiti-earthquake-where-did-money-go.

Reghezza-Zitt, M. (2013). *Utiliser la polysémie de la résilience pour comprendre les différentes approches du risque et leur possible articulation.* EchoGéo [En Ligne], 24. https://doi.org/10.4000/echogeo.13401.

Reghezza-Zitt, M., & Rufat, S. (2015). *Resilience imperative: Uncertainty, risks and disasters.* London and Oxford: ISTE Press and Elsevier.

Reghezza-Zitt, M., Rufat, S., & Djament-Tran, G. (2012). *What resilience is not: Uses and abuses.* Retrieved from http://cybergeo.revues.org/25554.

Reinhold, R., Navarro, M., & Rabinovitz, J. (1989). *The freeway dead: Portraits from Oakland—A special report. 11 whose lives ended as quake crushed I-880.* Retrieved from http://www.nytimes.com/1989/10/29/us/freeway-dead-portraits-oakland-special-report-11-whose-lives-ended-quake-crushed.html?pagewanted=all&src=pm.

Reisner, M. (2003). *A dangerous place*. New York: Pantheon Books.

Revet, S. (2007). *Anthropologie d'une catastrophe. Les coulées de boue de 1999 au Venezuela*. Paris: Presses Sorbonne Nouvelle.

Riesch, H., & Nowotny, H. (2017). Interdisciplinarity reloaded? Drawing Lessons from "citizen science". In B. Prainsack, S. Frickel, & M. Albert (Eds.), *Investigating interdisciplinary collaboration* (pp. 194–212). New Brunswick, NJ: Rutgers University Press. Retrieved from http://www.jstor.org/stable/j.ctt1j68m9r.14.

Rosbrow, T. (2012). Murakami's after the quake—The writer as waking dreamer and trauma analyst. *Psychoanalytic Dialogues: The International Journal of Relational Perspectives, 22*(2), 215–227.

Routley, G. J. (1991). *The East Bay Hills fire Oakland-Berkeley, California USFA-TR-060*. Oakland-Berkeley, CA. Retrieved from http://www.usfa.fema.gov/downloads/pdf/publications/tr-060.pdf.

Sanders, R. (2011). *Third modest quake within week sends message: Be prepared!* Retrieved from http://newscenter.berkeley.edu/2011/10/27/third-modest-quake-within-week-sends-message-be-prepared/.

Save San Francisco Bay. (n.d.). *Don't pave my Bay*. Retrieved from http://www.savesfbay.org/dont-pave-my-bay/history.

Schenk, G. J. (2007). Historical disaster research, state of research, concepts, methods and case study. *Historical Social Research, 32*(3), 9–31.

Schiewe, J. (2011, October 19). *After Oakland Hills fire, residents build of-the-wall homes*. Oakland North. Retrieved from http://oaklandnorth.net/2011/10/19/after-oakland-hills-fire-residents-build-off-the-wall-homes/.

Schoneker, J. (2010, January 24). *After the quake, Oakland's Haitian community reaches out to help*. Oakland North. Retrieved from http://oaklandnorth.net/2010/01/24/after-the-quake-oaklands-haitian-community-reaches-out-to-help/.

Schwartz, R. (2009, December). Eccentrics, heroes and cutthroats of old Berkeley. *SF Gate*. Retrieved from http://www.sfgate.com/bayarea/article/Eccentrics-Heroes-and-Cutthroats-of-Old-Berkeley-2523725.php#photo-2060993.

Seidman, S. (2008). *Contested knowledge: Social theory today* (4th ed.). Oxford: Wiley-Blackwell.

Self, R. (2005). *American Babylon: Race and the struggle for postwar Oakland*. Princeton: Princeton University Press.

Selin, C., Rawlings, K. C., de Ridder-Vignone, K., Sadowski, J., Altamirano Allende, C., Gano, G., … Guston, D. H. (2016). Experiments in engagement: Designing public engagement with science and technology for capacity building. *Public Understanding of Science*. https://doi.org/10.1177/0963662515620970.

Serres, M. (2006). *Homo pontifex. L'art des ponts*. Paris: Le Pommier.
Serres, M. (2009). *Temps des crises* (Manifestes). Paris: Le Pommier.
Shapin, S., & Shaffer, S. (2011). *Leviathan and the air-pump: Hobbes, Boyle, and the experimental life*. Princeton: Princeton University Press.
Shineha, R., & Tanaka, M. (2017). Deprivation of media attention by Fukushima Daiichi nuclear accident: Comparison between national and local newspapers. In J. Ahn, F. Guarnieri, & K. Furuta (Eds.), *Resilience: A new paradigm of nuclear safety: From accident mitigation to resilient society facing extreme situations* (pp. 111–125). Heidelberg and New York: Springer. https://doi.org/10.1007/978-3-319-58768-4.
Slater, D., Keiko, N., & Kindstrand, L. (2012). Social media, information and political activism in Japan's 3.11 crisis. *The Asia-Pacific Journal: Japan Focus, 10*(24). Retrieved August 12, 2019, from https://apjjf.org/2012/10/24/David-H.-Slater/3762/article.html.
Society, S. H. (2006). Palo Alto, Stanford, and the 1906 earthquake. *Sandstone & Tile, 30*(1), 40.
Solnit, R. (2009). *A paradise built in hell: The extraordinary communities that arise in disaster*. London: Viking.
Solnit, R. (2010). *Infinite city: A San Fransosco atlas*. Berkeley: University of California Press.
Sontag, S. (2004). *Regarding the pain of others*. London: Picador.
Sontag, D. (2012). Rebuilding Haiti lags after billions in post-quake AID. *The New York Times*. Retrieved from http://www.nytimes.com/2012/12/24/world/americas/in-aiding-quake-battered-haiti-lofty-hopes-and-hard-truths.html?ref=deborahsontag&gwh=F8EE9544A4C3DF90794C2960C75BDD13.
Souriau, E. (2009). *Les différents modes d'existence*. Paris: Presses Universitaires de France.
SPUR. (2013). *On solid ground*. San Francisco. Retrieved from http://www.spur.org/sites/default/files/publications_pdfs/SPUR_On_Solid_Ground.pdf.
Stallings, R. A. (1995). *Promoting risk: Constructing the earthquake threat (social problems and social issues)*. New York: Aldine de Gruyter.
Star, S. L., & Griesemer, J. (1989). Institutional ecology, "translation", and boundary objects: Amateurs and professionals in Berkeley's museum of vertebrate zoology, 1907–1939. *Social Studies of Sciences, 19*(3), 387–420.
Stavo-Debauge, J. (2004). Le pragmatisme et son public à l'épreuve du terrain. Penser avec Dewey contreDewey. In L. Quere & B. Karensti (Eds.), *Raisons Pratiques, 15* (Editions d, Vol. 15). Paris: Editions de l'EHESS.
Steinbrugge, K. (1968). *Earthquake hazard in the San Francisco Bay Area: A Continuing problem in public policy*. Berkeley: Institute of Governmental Studies, University of California.
Stengers, I. (1999). Savoirs Scolaires : qu'est-ce qui est digne d'être enseigné? In Conseil de l'Europe (Ed.), *La responsabilité, du principe aux pratiques Delphes*.

Retrieved from https://books.google.ch/books?id=BP5dhvqJ9TIC&pg=PA95&lpg=PA95&dq=savoirs+vivant++isabelle+stengers&source=bl&ots=1qsblPBk2S&sig=ACfU3U2FjwSthHnhkBfCUTkq4DppJJ7w-6w&hl=en&sa=X&ved=2ahUKEwiclrSNhd_hAhVU8aYKHfwuDyQQ6AEwCHoECAkQAQ#v=onepage&q=savoirs vivant isabelle stengers&f=false.

Stengers, I. (2002). Un engagement pour le possible. *Cosmopolitiques, 1*, 27–36.

Stengers, I. (2009). *Au temps des catastrophes; résister a la barbarie qui vient* (les empêch). Paris: La Découverte.

Stengers, I. (2015). *In catastrophic times: Resisting the coming barbarism* (Critical C). London, UK: Open Humanities Press and meson press. Retrieved from http://openhumanitiespress.org/books/download/Stengers_2015_In-Catastrophic-Times.pdf.

Stengers, I., & Latour, B. (2009). Le sphinx de l'oeuvre. In *Les différents modes d'existence: Suivi de Du mode d'existence de l'oeuvre à faire* (pp. 1–75). Paris: Presses Universitaires de France.

Stock, M. (2004). *L'habiter comme pratique des lieux géographiques.* Retrieved from http://www.espacestemps.net/articles/lrsquohabiter-comme-pratique-des-lieux-geographiques/.

Strauss, J. A., & Allen, R. M. (2016). Benefits and costs of earthquake early warning. *Seismological Research Letters, 87*(3), 765–772. https://doi.org/10.1785/0220150149.

Strong, T. (2008). A review of Lorraine Daston and Peter Galison's objectivity. *The Weekly Qualitative Report, 1*(10), 62–66.

Sunstein, C. (2002). *Hazardous heuristic.* The Law School of the University of Chicago, Public Law. Retrieved from http://www.law.uchicago.edu/files/files/33.crs_.hazardous.pdf.

Taira, T., Silver, P. G., Niu, F., & Nadeau, R. M. (2009). Remote triggering of fault-strength changes on the San Andreas fault at Parkfield. *Nature, 461*(7264), 636–639. https://doi.org/10.1038/nature08395.

Taussig, M. (2009). Culture of terror—Space of death: Roger Casement's Putumayo report and the explanation of torture. *Comparative Studies in Society and History, 26*(3), 467. https://doi.org/10.1017/S0010417500011105.

Taylor, T. (2011). 20 years after the fire, revelation and tribute. *The New York Time.* Retrieved from http://www.nytimes.com/2011/10/21/us/richard-misrach-reveals-his-images-of-oakland-berkeley-fire.html?_r=0.

Taylor, W. M. (2006). Urban disasters: Visualising the fall of cities and the forming of human values. *Journal of Architecture, 11*(5), 603–612.

Thatcher, W., Ward, P., Wald, D., Hendley, J. I., & Stauffer, P. (2001). *When will the next great quake strike Northern California?*

The Post and Courrier with A.P. (1992). Town Gets Quake Warning. *The Post and Courier.* Retrieved from http://news.google.com/newspapers?nid=2482&dat=19921021&id=wlZSAAAAIBAJ&sjid=1DYNAAAAIBAJ&pg=6194,1792815.

Thien, D. (2005). After or beyond feeling? A consideration of affect and emotion in geography. *Area, 37*(4), 450–454. https://doi.org/10.1111/j.1475-4762.2005.00643a.x.

Thrift, N. (2008). *Non-representational theory, space, politics, affect (International Library of Sociology)* (J. Urry, Ed.). London: Routledge—Taylor & Francis Group.

Tierney, K. J. (1994). *Sociology unique contributions to the study of risk* (No. #204).

Tierney, K. J. (2000). *Public support and priorities for seismic rehabilitation in the East Bay region of Northern California* (No. #296).

Tierney, K. J. (2001). *How will social science help us deal with earthquake?* Chicago: University of Chicago Press.

Timmerman, P. (1981). Vulnerability, resilience and the collapse of society: A review of models and possible climatic applications. *Environmental Monograph*. https://doi.org/10.1002/joc.3370010412.

Tobriner, S. (2006). *Bracing for disaster. Earthquake-resistant architecture and engineering in San Francisco, 1838–1933.* Berkeley: The Bancroft Library, University of California Berkeley.

Traweek, S. (2013). *Downwind and denial: An STS forum on Fukushima.* Retrieved from https://fukushimaforum.wordpress.com/workshops/sts-forum-on-the-2011-fukushima-east-japan-disaster/manuscripts/session-3-radiation-information-and-control/downwind-and-denial/.

Tulloch, J., & Lupton, D. (2003). *Risk and everyday life.* London: Sage.

Turner, F. (2004). Seventy years of the riley act and its effect on California's building stock. In *13th World Conference on Earthquake Engineering*. Vancouver.

University of Buffalo. (2008). *Jack London and the 1906 earthquake.* Retrieved from http://mceer.buffalo.edu/1906_Earthquake/personal_stories/jack-london.asp.

USGS. (2013). *Are Earthquakes really on the increase?*

U.S. Geological Survey. (2015). *M6 South Napa, California earthquake—August 24, 2014.* Retrieved November 2, 2019, from https://earthquake.usgs.gov/earthquakes/events/2014napa/.

U.S. Geological Survey. (2016). *When could the next large earthquake occur along the San Andreas fault?*

Vincent, C. (2017, December 29). En 2018, vous ne verrez plus le réchauffement climatique comme une fatalité grâce à la philosophe Emilie Hache. *Le Monde*. Retrieved from http://www.lemonde.fr/idees/article/2017/12/29/en-2018-vous-ne-verrez-plus-le-rechauffement-climatique-comme-une-fatalite-grace-a-la-philosophe-emilie-hache_5235552_3232.html#vZtFGWSXHFPtoISU.99.

Von Winterfieldt, D., Roselund, N., & Kisuse, A. (2000). *Framing earthquake retrofitting decisions: The case of hillside home in Los Angeles.* Pacific Earthquake

Engineering Research Center. Retrieved from http://peer.berkeley.edu/publications/peer_reports/reports_2000/0003.pdf.

VonWinterfeldt, D., Roselund, N., & Kitsuse, A. (2000). *Framing earthquake retrofitting decisions: The case of hillside homes in Los Angeles.*

Wald, D. (2001). *Did you feel it? Community-made earthquake shaking maps.*

Walker, R. (1990). The playground for US capitalism? The political economy of San Francisco Bay Area in the 80's. In M. Davis, S. Hiatt, M. Kennedy, S. Ruddick, & M. Sprinker (Eds.), *Fire in hearth: The radical politics of place in America.* London: Verso.

Walker, R. (1995). Landscape and city life: Four ecologies of residence in San Francisco Bay Area. *Cultural Geographies, 2*(1), 33–64. Retrieved from http://oldweb.geog.berkeley.edu/ProjectsResources/Publications/BayLandscape.htm.

Walker, R. (1998). An Appetite for the City. In J. Brook, C. Carlsson, & N. J. Peters (Eds.), *Reclaiming San Francisco, history, politics, culture* (p. 355). San Francisco: A City Lights Anthology.

Walker, R., & Thomas, S. (2010). Blinded by history: The geographic dimension of environment and society. In D. Cazaux Sackman (Ed.), *A companion to american environmental history* (pp. 553–578). Blackwell Publishing. https://doi.org/10.1002/9781444323610.

Walter, F. (2008). *Catastrophes. Une histoire culturelle XVI-XXIème siècle.* Paris: Editions du Seuil.

Wang, A. (2011, October). Earthquake Predictions trigger anxiety, fear in Berkeley community. *The Daily Californian.*

Watts, M. J., & Bohle, H. G. (1993). The space of vulnerability: The causal structure of hunger and famine. *Progress in Human Geography, 17*(43), 43–67. https://doi.org/10.1177/030913259301700103.

Weichselgartner, J., & Kelman, I. (2015). Geographies of resilience: Challenges and opportunities of a descriptive concept. *Progress in Human Geography, 39*(3), 249–267. https://doi.org/10.1177/0309132513518834.

Whatmore, S. J. (2009). Mapping knowledge controversies: Science, democracy and the redistribution of expertise. *Progress in Human Geography, 33*(5), 587–598. https://doi.org/10.1177/0309132509339841.

White, G. (1945). *Human adjustment to floods—A geographical approach to the flood problem in the United-States.* The University of Chicago. Retrieved from http://www.colorado.edu/hazards/gfw/images/Human_Adj_Floods.pdf.

Whitney, L. (2013). *Google Person Finder helping to find missing in Philippine floods.* C/Net.

Wiener, J. B. (2010). *The real pattern of precaution.* Washington, DC: Future Press.

Wiener, J. B., & Rogers, M. D. (2002). Comparing precaution in the United States and Europe. *Journal of Risk Research, 5*(4), 317–349. https://doi.org/10.1080/1366987021015368.

Wikipedia Contributors. (2018a). Henry J. Degenkolb. In *Wikipedia, The Free Encyclopedia*. Wikipedia, The Free Encyclopedia. Retrieved from https://en.wikipedia.org/wiki/Henry_J._Degenkolb.

Wikipedia Contributors. (2018b). Stafford Disaster Relief and Emergency Assistance Act. In *Wikipedia, The Free Encyclopedia*. Wikipedia, The Free Encyclopedia. Retrieved from https://en.wikipedia.org/w/index.php?title=Stafford_Disaster_Relief_and_Emergency_Assistance_Act&oldid=827132167.

Wikipedia Contributors. (2018c). Unreinforced masonry building. In *Wikipedia, The Free Encyclopedia*. Retrieved from http://en.wikipedia.org/wiki/Unreinforced_masonry_building.

Wikipedia Contributors. (2019a). Bruce Bolt. In *Wikipedia, The Free Encyclopedia*. Wikipedia, The Free Encyclopedia. Retrieved from https://en.wikipedia.org/wiki/Bruce_Bolt.

Wikipedia Contributors. (2019b). Prospect theory. In *Wikipedia, The Free Encyclopedia*. Wikipedia, The Free Encyclopedia. Retrieved from http://en.wikipedia.org/wiki/Prospect_theory.

Wikipedia Contributors. (2019c). Soft story building. Retrieved May 15, 2019, from https://en.wikipedia.org/w/index.php?title=Soft_story_building&oldid=885792427.

Winchester, S. (2006). *The Crack in the edge of the world: America and the great California earthquake of 1906*. New York: Harper Perennial.

Winchester, S. (2011, March 13). The scariest earthquake is yet to come. *Newsweek Magazine*. Retrieved from http://www.thedailybeast.com/newsweek/2011/03/13/the-scariest-earthquake-is-yet-to-come.html.

Wisner, B., Blaikie, P., Cannon, T., & Davis, I. (2006). *At risk*. London: Routledge.

Working Group on California Earthquake Probabilities. (1999). *Earthquake probabilities in the San Francisco Bay region : 2000 to 2030—A summary of findings. Open-file report 99-517*. Sacramento, CA: U.S. Geological Survey and US Departement of Interior. Retrieved from https://pubs.usgs.gov/of/1999/0517/pdf/of99-517.pdf.

Working Group on California Earthquake Probabilities. (2003). *Earthquake probabilities in the San Francisco Bay region: 2002–2031. Open-file report 03-214*. U.S. Geological Survey. Retrieved from https://pubs.usgs.gov/of/2003/of03-214/.

Wynne, B. (1996). May the sheep safely graze? A reflexive view of the expert-lay knowledge divide. In S. Lash, B. Szersynsky, & B. Wynne (Eds.), *Environment and modernity: Toward a new ecology* (pp. 27–44). New York: Sage.

Yamada, K. (2012). *The Bay Area's response to Japan's 3-11 catastrophe*. Retrieved from http://japancenter.sfsu.edu/sites/sites7.sfsu.edu.japancenter/files/uploads/HowHopeSpringsfromDisaster.pdf.

Index

C. Mazel-Cabasse, *Waiting for the Big One*,
https://doi.org/10.1007/978-3-030-15289-5

R

S

T

Kolibri 360 GmbH
Ferdinand-Jühlke-Straße 7
99095 Erfurt, Deutschland
produktsicherheit@kolibri360.de